AF568087

ENVIRONMENT EVERYONE

BITTER-SWEET WRITINGS ON ENVIRONMENT

ENVIRONMENT EVERYONE

BITTER-SWEET WRITINGS ON ENVIRONMENT

PROFESSOR S. A. ABBASI,
PhD, DSc, FIE, PE
Senior Professor & Director
Centre for Pollution Control & Energy Technology
Pondicherry University,
Pondicherry 605 014

1998

Discovery Publishing House

New Delhi-110002

First Published-1998

ISBN 81-7141-439-7

Published by:

Discovery Publishing House

4831/24, Ansari Road, Prahlad Street,
Daryaganj, New Delhi-110 002 (*INDIA*)
Phone: 3279245
Fax: 91-11-3253475

Printed at:
Arora Offset Press
Laxmi Nagar, Delhi-92

Dedicated to

BROTHER DR. KARAN KUMAR SINGH BHATIA,
NEERU BHABHI,
KINNARI and PALLAVI

CONTENTS

FOREWORD

Prof. Abbasi's work of compilation, captioned 'Environment Everyone' brings together his writings on various aspects of environmental protection and ecological balance.

In modern world, the concept of development needs to be re-oriented in the futuristic parameters of its sustainable condition. Over the past decades, the significance of environment movement has grown to enormous proportions in almost all parts of the world. The depletion of atmosphere ozone layer and climatic imbalances caused by emission of 'greenhouse' gases have unleashed grim challenges to environmental and ecological concerns of humankind. As has been rightly pointed out by the author, the environment has become a serious concern of greater consequences to the developing nations such as is ours with tropical climate, burgeoning population and poverty than to the developed nations. While the author delved deep into the fascinating features of Nature's gifts such as flowers and water, he has also recounted some tragic stories such as 'Minamata' which should serve as an eye-opener for the society to commit itself with firm determination to uphold the worthy cause of protecting environment at all costs.

I am sure that 'Environment Everyone' will be a veritable source of useful information for its readers who evince keen interest in the study of matters pertaining to environment.

Rajani Rai

RAJANI RAI
Lieutenant-Governor

PONDICHERRY

PREFACE

Environment may mean different thing to different people but it invariably means *something* to everyone. And environment is everywhere: air environment, water environment, land environment, business environment, political environment, cultural environment, claustrophobic environment, colourful environment.....

The author has been with the environment movement right from the time it all began in the late 1960s. He was a part of the brigade which patted itself on its back for being *avant* garde, of being 'in' with the *nouveau vogue.* He was also among those who faced redicule for the same reason; who were accused of aping the West. "Environmental pollution is the new fad of the developed countries", the critics used to thunder, "like McDonalds, computerised dating, and Chubby Chekker.....What relevance does it have for a poor nation like India? We don't have food to eat..." etc.

Well, the times have a-changed. And fast. We know that environment is perhaps a greater concern to us than to the developed nations. With our tropical climate, burgeoning population, and poverty, WE are likely to be hit a lot harder by environmental degradation than THEY.

Environement Everyone is a collection of essays I have written over the years for the audio-visual and print media. It covers a wide range of topics, generates feelings varying from despair to hope, and leaves tastes ranging from bitter to sweet.

There are intriguing questions-such as about India's most harmful pollutant and the most important animal on earth-with attempted answers. There are true-life stories *(Minemata)* and memories, and there are write-ups looking at the science of beauty *(Flowers)*. Different colours, different odours, different sounds, different tastes-just as we find in the 'environment'. I have absolutely no pretention that I have captured all, or even a substantial number of the innumerable dimensions of environment.

At best I have managed a sample.

S. A. ABBASI

PART -1

ESSAYS

1

INDIA`S MOST HARMFUL POLLUTANT

Which is the source of the most harmful pollution in India; pollution which causes maximum harm in terms of number of people it effects?

Pulp & paper industries? No.

Pesticide industries? No.

Thermal power plants? No.

Neculer reactors? No.

It is natural to think of industries as soon as we hear the word 'pollution'. Industries *are* the most conspicuous of all polluters. This is because industries are situated in or near big cities where major newspapers, TV, and radio stations are also situated. So are major journalists. So are articulate citizens, law-makers, and law-breakers.

All this contributes towards the public attention that is attracted by industries *vis a vis* environmental pollution.

But no single industry or type of industries qualify as the most harmful pollutor in India. Even all industries *combined* do not qualify!

MOST DAMAGING POLLUTANTS

The most harmful and the most damaging pollutants, from the point of view of number of deaths they cause every year and number of people they render partially or fully ineffective every year, are let out by us! Us humans!

Yes. The nightsoil, sewage, and garbage generated by us Indians are the biggest, the most widespread, and the most damaging pollutants that are released in India!

Indeed it is so in other developing countries as well!

We do hear, now and then, about episodes of damage to public health occurring due to industrial pollution. An accidental release of rayon effluents in Bombay toxifies a sewer ... a bonefire involving some toxic wastes chokes a neighbourhood in Delhi ... cases of knock-knee desease occur in Andra Pradesh due to molybdenum poisoning ... the frequency of respiratory ailments increases due to fly ash ...

Disconcerting as these reports are, worth attention as these reports are, they are never able to match the frequency of the reports we get of ourbreaks of cholera, jaundice, dysentery, typhoid, maleria, yellow-fever ... Except the Bhopal gas disaster which was caused by *accidental leak* and not day-to-day pollution no other episode of industrial pollution can match the national panic triggered by the rats of Surat. Those rats were attracted, nurtured, and infected by nothing else but the garbage which the citizens of Surat had generated; and had done so out of proportion to the capacities of the city's civic amenities to handle that garbage.

WATER AND WATERBORNE DISEASES

In all developing countries, ill health is largely due to lack of safe drinking water. Statistics show that disease rates and mortality can drastically be cut down by the imporovement of quality of public water supply. For example, a study conducted in Uttar Pradesh has revealed that after protected water supply was introduced, cholera deaths decreased by 76.1%, typhoid fever death rate by 63.6%, dysentery death rate by 23% and the rate for diarrhoeal diseases by 42.7%. Therefore the resources spent on water supply schemes can yield huge dividends in terms of improved health - consequently improved productivity. A survey by WHO (World Health Organisation) on community water supplies has revealed that in India while three-fourth of the population in urban areas has access to community water supplies, only one-fourth of the rural population enjoys this basic amenity. The slow pace of progress in this matter is due to a number of factors. Some are :

i) Lack of funds
ii) Shortage of trained manpower
iii) Weakness in the implementation of water supply programmes in terms of :
    a) Difficulties in system operation and maintenance
    b) Insufficient involvement of potential users

In 1976 United Nations Conference on Human Settlements adopted the resolution that they will make global efforts to achieve the target of 'water for all by 1990'. Well, 1990 has come and gone buttheWorldis waybehind thetarget.

USES OF WATER

Water is used mainly for four purposes :

i) Domestic - drinking, cooking, washing, bathing.
ii) Public - public cleaning, fire-fighting, maïntenance of public gardens, swimming pools and other civic amenities.
iii) Industrial
iv) Agricultural

Water for domestic use has to be 'safe' and 'wholesome'. Safe water means that it cannot harm the consumer even when used for long periods. A water may be safe, but if it has an unplesant taste or appearance, it may drive the consumer to other less safe sources. Therefore, drinking water has to be agreeable to use and wholesome, it should be *aesthetically acceptable* as well as *potable*. Safe and wholesome water should be :

a) free from pathogenic agents,
b) free from harmful chemical substances,
c) pleasant to the taste,
d) usable for domestic purposes.

Water may be contaminated or polluted by infective and parasitic agents, poisonous chemicals, industrial or other wastes or sewage.

WATER POLLUTION

Impurities in water are dissolved or suspended, such as :

i) dissolved gases,like hydrogen sulfide, carbondioxide, ammonia, nitrogen;

ii) dissolved minerals, such as, calcium, magnesium and sodium salts;

iii) suspended impurities, such as, clay, salt, sand, mud;

iv) microscopic plants and animals;

v) industrial and trade wastes containing toxic agents ranging from metal salts to conplex synthetic organic chemicals;

vi) agricultural pollutants, such as, fertilizers and pesticides;

vii) physical pollutants, such as, thermal pollutants and radioactive substances.

DISEASES RELATED TO WATER

Water-borne diseases have two origins :

i) biological

ii) chemical.

Biological sources may include :

A. *INFECTIVE AGENTS* :

a) viral; causing viral hepatitis, poliomyelities, swimming pool conjuctivitis, viral gastroenteritis;

b) bacterial; causing cholera, typhoid, paratyphoid, bacillary, dysentery, gastroenteritis, infantile diarrhoea, tularaemias;

c) protozoal; causing amoebiasis, giardiasis;

d) helminthic; causing roundworms, whipworm, threadworm, hydatid disease;

e) leptospiral; causing weil's disease;

f) fungal; causing rhinisporidiosis.

B. *AQUATIC HOSTS* :

Cyclops, guinea-worms, fish tapeworms, Snail, Schistosomiasis.

C. *BREEDING OF VECTORS* :

Causing mosquito-borne diseases such as filaria and malaria, dengue fever, Japanese B. encephalities. Chemical sources may consist of :

A. Industrial and agricultural wastes acting as *pollutants*:

detergents, solvents, cyanides, heavy metals, minerals, organic acids, nitrogenous substances, bleaching agents, dyes, pigments, sulfides, toxic biocidal agents.

B. *Excess of fluorine* - causing fluorosis.

C. *Deficiency of iodine* - causing endemic goitre.

Endemic fluorosis is produced by excess of fluorine in drinking water. One form of fluorosis is dental fluorosis producing discolouration and mottling of teeth. The worst form of the disease is skeletal fluorosis which results in pain and stiffness of joints and spine and finally bed-ridden stage. Fluorosis is mostly reported from certain districts of Andra Pradesh and Punjab. It has also been reported in Kanyakumari.

The iodine deficiency which produces goitre, mental defects, nervous defects and cretinism can originate from drinking water deficient in iodine. The endemicity of iodine deficiency in the Himalayan ranges is well documented. It is also common in Kerala, specially ldukky, Thodupuzha and other hilly tracts. More community based surveys and analysis of drinking water for iodine content is required. Iodine deficiency can be prevented by iodisation of salt or by the injection of iodised oil. Persons suffering from goitre are under a high risk of developing cancer of the thyroid gland.

Pollution of water by industrial effluents is increasingly becoming a problem in India next in scale only to the problems due to domestic wastes. The episodes of mercury poisoning producng Minimata disease and cadmium poisoning leading to *itai itai* disease in Japan are well known by now. In India there have been reports of molybdenum poisoning in Andhra Pradesh - leading to Knock-Knee disease called 'genu velgam' in Telugu. Industrial effluents can adversely affect the flora and fauna and indirectly affect the human existence.

WATERBORNE DISEASES IN INDIA-WITH SPECIAL REFERENCE TO KERALA

In India health hazards related to drinking water are predominantly biological in origin, *viz.*, viral, bacterial, protozoal and helminthic disorders caused by the contamination of water by human refuse. Eventhough there have been sporadic (and serious) cases of drinking water pollution due to geological contamination -the incidence of arsenic poisoning in West Bengal being a well-known recent example - such cases are much more rare than the ones involving biological pollution. Table 1 presents an illustrative example of water borne diseases, with reference to Kerala.

Most of these are due to faecal contamination of drinking water. The prevalence of these disorders do not appear to be geographically distributed and are seen to occur in coastal, mid-land and high-land areas alike. There is a seasonal prevalance for cholera during the months of June, July, August and September (Table 2).

However, viral hepatitis is reported throughout the year, with greater frequency in July, August and September (Table 3). It may be noted that these are only illustrative data from Directorate of Health Services. Some 90% of the viral hepatitic cases go to Ayurveda and overall 60% cases go to private practitioners which do not come in the Directorate's data.

Enteric fever is also reported throughout the year (Table 4).

Incidences of cholera have decreased from what it was two decades ago. However the fact that cholera is still reported from coastal districts during the monsoon poses a major public health threat. Malaria is at present not endemic in Kerala.

However about four thousand cases are being reported annually. There seem to be some imported cases as well (Table 5).

Table 1.

Morbidity pattern as seen in Primary Health Centres in Kerala (as percentage of total outpatient attendence)

Kulangara	*Disease*	*Poovar*	*Vilap- Anjengo*	*Kanniya- pil sala*
Resp. infections	28.2	35.2	28.8	23.1
Diarrhoeal disorders	8.2	13.6	5.1	9.7
Skin infections	20.6	8.0	7.5	8.7
Fevers other than				
enteric fever	2.4	4.8	9.4	7.2
Helminthiasis	3.5	.	4.7	7.7
Injuries	5.3	10.4	3.5	3.1
All other diseases	32.2	28.0	41.5	40.5

Table 2.

Incidence of cholera in Kerala

	No of cases	
Year	*June - September*	*November - May*
1981	28	5
1982	21	7
1983	209	17

Table 3.

Incidence of viral hepatitis in Kerala

Year	*No. of cases*	*No. of deaths*
1981	21127	45
1982	20215	27
1983	18927	14

With water as well as the vector mosquitoe available in plenty in Kerala, the risk of malaria assuming epidemic proportions is very high. Filariasis is rampant in the state all along the coastal districts and even in certain pocketsin the foothills of the Western ghats. A survey revealed that in 1960s a total of 13,45,861 people were positive for microfilaria in the blood; the maximum prevalance being in Alleppey, Cannanore, Ernakulam and Malappuram Districts. Sadly this only may be the proverbial tip of the iceberg and there may be thousands of more people with one or other overt manifestation of filarial infection. The population that is exposed to this infection obviously is very huge and without mosquito control measures, the incidence is likely to go up. If this thirteen lakhs and odd number of individuals are to be provided with prophylatic specific drug, namely diethyl carbamazine for a period of two weeks, the economic drain on the government will be considerable. For the eradication of the disease the same carrier may require repeat does of the drug. Surveillance measures and mosquito control measures including the removal of water hyacyinth, and pistia plants from the ponds and canals and waterlogged areas will consume huge amounts.

Table 4.

Incidence of enteric fever in Kerala

Year	*No. of cases*	*No. of deaths*
1981	13253	18
1982	13138	8
1983	7515	0

Table 5.

Incidence of malaria in Kerala

Year	*No. of cases*	*No. of deaths*
1981	4127	0
1982	3981	0
1983	3725	0

Rhinosporidiosis is a fungus infection predominantly affecting the upper respiratory passages like nose and throat transmitted by bathing in ponds, and in tanks contaminated by the spores, prevalant in the coastal districts with over hundred patients admitted to Medical College Hospital (MCH) per annum. Surgery is required for eradication and chronic morbidity can be caused (Table 6).

Poliomyelitis is a major cause for childhood morbidity and mortality and its causative virus is spread through food and water. Figures from SAT hospital, Trivandrum, demonstrate a rising trend in the incidence in the recent years (Table 7).

Numbers denote number of patients for every 1000 medical paediatric admissions.

Exact incidence of this illness like that of viral hepatitis is difficult to be assessed because of the fact that most patients go to the local ayurvedic physician or quacks in the belief that it is a type of rheumatism. Western countries have almost totally conquered this crippling disease in children by total immunisation of the new borns, and health education. The degree of handicap due to this disease varies from paralysis of one limb to paralysis of all the limbs. Thus if dysentry, diarrhoea, cholera and gastroenteritis take away on the average seven to ten working days per persons, viral hepatitis and enteric fever take on an average three weeks if not more for total recovery and for the patient to be fit enough to resume normal rigours of work. Viral heptitis in addition is capable of producing prolonged morbidity by way of complications like chronic hepatitis.

Table 6.

Reports of rhinosporidiosis in MCH, Thiruvanathapuram

Year	*No. of cases*
1980	36
1981	144
1982	115
1983	110

Table 7.

Proporation of children treated for selected communicable diseases in S.A.T. HOSPITAL 1964-80 (number per 1000 patients admitted)

	64-67	67-70	70-73	73-76	80
Anterirorpoliomyelitis	11	7	12	15	20
Gastroenteritis	198	201	209	210	201
Dysentery	64	37	32	27	21
Cholera	0	0	0	0	0
Viral hepatitis	39	37	37	24	31
Enteric fever	30	30	32	21	5
Total	342	318	322	297	278

Gastroenteritis in infants and children is very often the precipitating factor for the onset of more moribund states like protein-calorie malnutrition and flaring up of latent tuberculosis. In the nutritionally imbalanced paediatric age group, intercurrent infections of the gastro intestinal tract tilts the balance to the weaker side. It is well known that an epidemic of gastroenteritis is followed in a month or two by an epidemic of Kwashiorkor or protein calorie malnutrition. Helminthiasis like ascariasis trichuriasus and enterobiasisi are indirectly related to poor personal hygiene and use of water and adds to the misery caused by diseases like dysentery, diarrhoea, enteric fever, and gastroenteritis. In non-coastal villages only 10% of all age groups are free of any helmintic infection (Table 8). As much as 80% of the population have ascariasisis. A survey at SAT Hospital, Trivandrum, showed that over 90% of the children admitted for treatment harboured helminthic infection irrespective of the menifest disease they came for (Table 9). It has been estimated that on an average 300 out of every thousand children admitted suffer from a water borne disease.

The number of work-days lost and the number of work-days on which a person performs at less than his/her 100% efficiency due to morbidity caused by waterborne diseases are enormous (Table 10). In terms of loss of man-days and cost of treatment the consequent drain on the Nation's economy would work out to be several thousand crores of rupees per annum.

Table 8.

Helminthic Infestation pattern in a noncoastal village Cenniyoor (percent of persons in each age group affected by each worm)

Age in years	*Round worm*	*Hook worm*	*Whip worm*	*Non infected*
1 - 5	78.8	41.4	33.7	-
6 - 10	85.5	45.0	42.2	-
11 - 10	75.0	47.0	47.6	-
16 - 25	81.5	41.6	30.6	-
26 - 35	85.5	50.4	40.6	-
36 - 45	81.4	48.9	48.8	-
45 & above	75	45.3	49.0	-
All age groups	80.2	43.6	40.3	10.2

Table 9.

Helminthic infestation pattern among institutionalised children aged 5-17 years in three natural divisions of Kerala (percent of persons)

Stool examination	*Highlands*	*Midlands*	*Coastal*
Round worm	72	77	76
Hook worm	26	2	24
Whip worm	15	15	22
Mixed infection	32	17	28
Total Positive Stools	100	92	92

Table 10.

Monetary losses due to waterborne diseases in Kerala

Disease	*Average* No.of *cases*	*Percapita* duration *of ill ness*	*Work* expenditure *per day*	*Total* days *lost*	cost *Rs.*
Dysentery	1021842	7	Rs 10	7152894	1,02,18,420
Diarrhoea	19613	7	10	137291	1,96,130
Gastroenteritis	17507	7	20	122549	3,50,140
Enteric fever	14138	21	100	296898	14,13,800
Viral hepatitis	20215	21	50	424515	10,10,750
Cholera	21	7	25	147	525
Malaria	3981	10	5	398110	19,905
Total	..	..	..	8174104	1,32,09,670

2

THE MINAMATA STORY

The episode which brought the attention of the world towards heavy metal pollution in most poignant and dramatic manner, occurred forty years ago - in 1956. In April of that year the first case of what subsequently came to be known as *Minamata Disease* was registered. A six-year old girl was admitted to the hospital of The New Japan Chisso Fertilizer Company, situated near Minamata Bay on the coast of the southern island of Kyushu in Japan. The girl had symptoms that resembled brain damage. She had difficulty in walking or talking normally and was at times delirious. Dr Hajime Hosokawa, who examined her, could not place her disease and termed it, 'an unclarified disease of the central nervous system'. Some months later, of course, the disease was established as mercury poisoning.

What came to surface in 1956 had its origin in 1908 and its victims continue to suffer even till today. What is worse is that the Minamata Disease has not yet claimed its last victim.

There are many many lessons to be learnt from Minamata episode; lessons which we would do well to learn for our own good as also of others. But first let us recapitulate what happened at Minamata.

THE MINAMATA DISEASE STORY

The New Japan Chisso Fertilizer Company was established in Minamata in 1908. It was perceived as a boon by the local population, of which the dominant occupation was fishing, as the factory was expected to provide better employment opportunities to them.

In 1932, the factory began manufacturing acetaldehyde, a process which requires, amongst other chemicals, methyl mercury chloride. The untreated waste from the factory was discharged in the nearby Minamata bay, regardless of the fact that local fishermen consumed vast quantities of fish and seafood caught from that area.

From 1950 onwards dead fish were found floating in the bay every now and then but this was not linked to the Chisso factory. Two years later, the cats in Minamata began to exhibit strange symptoms. They would stagger about as if drunk, salivate, have convulsions and suddenly collapse. The disease became so prevalant that it almost viped out the entire feline population in the area. All this while the local fishermen continued to fish in the bay and their catch formed a regular part of the diet of the 50,000 - strong local populace of Minamata.

In April 1956, the first case, described at the beginning of this write-up, of what later came to be known as *Minamata Disease* was registered. Around this time other people also began to show similar symptoms. The disease aquired the name *the cats dancing disease* or *the strange disease* among the local populace.

As there was no clear diagnosis of the cause of the disease, and often more than one member of each family, or people living in the same neighbourhood, were afflicted by the disease, it was assumed that the disease was contagious. This belief led to shunning and isolation of the victims.

In October 1956, the Minamata Disease Research Group of Kumamoto University Medical School made public its finding that the disease was not infections. It concluded that the disease was the result of heavy-metal poisoning caused by eating fish and shellfish from Minamata Bay. The specific metal, however, had not yet been identified. After these findings, the first warning against eating fish was issued. In other words, from 1950, when the first dead fish floated on the surface of the bay, till 1956, the people of Minamata continued to eat the local fish.

According to Tsuginori Hamamoto, who is one of the leaders of the Minamata Victim's Group, people continued to fear touching or being too close to victims of Minamata disease even after it was announced that the disease was not contagious. He said he had also tried to hide the fact that he had the disease and sought part-time work.

By this time there were 56 known victims of the disease. A year later, the number of dead had risen to 21. In 1958, the local government issued a ban on the sale of the fish but did not ban fishing in the bay. Meanwhile, Chisso began diverting its waste effluent from the bay to a nearby river.

A full three years after the first registered case of the Minamata Disease, the Kumamoto university research team was able

to establish that it was methyl mercury chloride in the Chisso factory's effluent which had caused the disease. This finding was further confirmed by Dr Hosokawa, the Chisso doctor, through experiments on cats. When the Chisso corporation knew of Dr Hosokawa's findings it told him categorically to discontinue his investigations and not to make his findings public.

What is worse, despite this information the Chisso company continued to dump mercury until 1968, when the Japanese government intervened. In 1959 it also successfully persuaded the first victims of mercury poisoning into entering into a settlement which would absolve the company of any further liability.

The people who settled were ignorant about the true nature of mercury poisoning, or the fact that it was congenital.

By the time it was conclusively established in 1962 that the methyl mercury chloride in the acetaldehyde sludge from the Chisso factory had made its way through the food chain into the bodies of people, leading to the Minamata disease, 121 victims had been verified of whom 46 were dead.

Even today, forty years after it all began, the issue refuses to die as new cases of Minamata disease continue to surface. Women who had consumed fish from the bay have given birth to children with the disease. In fact, its worst manifestations were in these children, numbering over 50.

Apart from verifying and treating the victims, the local Government had to undertake the expensive task of completely filling the Minamata Bay because traces of mercury - of which an estimated 600 tonnes had been dumped - could still be found on the sea bed. Rather than risk further contamination of fish and sea food or a permanent ban on fishing in the bay, the Government decided to reclaim a part of the bay through an enormous landfill.

The most heart - rendering aspect of Minamata episode is that even today there are people who are suffering from some aspects of mercury poisoning, but who have still not been recognised as victims. Because the earlier symptoms resembled brain damage, it was assumed that mercury intake leads primarily to damage of the nervous system, but later studies have established that the metal also affects other organs and can disturb liver function and cause diabetes and hypertension. It has also been established that the disease is congenital and is passed on to coming generations. However, in the absence of clear cause - effect links as in Minamata, children suffering from Minamata disease could be diagnosed as suffering from mental deficiency, and are thus kept outside the perifery of people eligible for compensation.

As of today over 3000 people in three prefectures have been recognised as Minamata disease victims of whom 1144 have been dead. Several thousand cases of people who had been examined but whose cases have not been conclusively linked with Minamata, or people who had yet to be examined, were pending.

THE LESSONS

We learn several important lessons from this story.

a) Heavy metal pollution can remain dormant for a long time and then surface with vengeance. Today in many areas heavy metal build-up is occurring due to disposal of industrial effluents. No direct impacts on human health have yet been seen. But this could well be the calm before the storm.

b) Once an area gets toxified with heavy metals it is extremely difficult - almost impossible - to detoxify it.

c) Symptoms of heavy metal toxicity are often similar to the symptoms of other common diseases - such as respiratory problems, digestive disorders, skin diseases, hypertension, diabetes, jaundice etc. This makes it all the more difficult to quickly diagnose such heavy metal poisoning. Sufferers can continue to be treated for similar - looking diseases till the problem reaches alarming proportions.

d) The damage caused by heavy metals to a person does not necessarily end with the life of that person. The harmful effects can be transferred to the person's progenies.

IN SUMMARY

Ironically heavy metal pollution is a direct offshoot of our increasing ability to mass-produce metals and use them in all spheres of existence. We seek to enhance the quality of our life by mastering the sciences of metallurgy and materials technology. In the process we also jeopardize the quality of our life by exposing ourselves to heavy metal pollution!

3

FLOWERS

Flowers are amongst those gifts of nature whose appeal transcends all boundaries of race, religion, region and culture. The sheer joy of beholding a flower can be paralleled with that of listening to music; only the former does not require any initiation whatsoever - a flower appeals to a child as much as it does to an old man; it appeals to an illiterate as much as it does to an erudite person. It is small wonder that in the literature, especially poetry of all languages across the world, flowers have inspired more musings than any other motif outside the domain of human relationships. Flowers have always been used to signify beauty, holiness, innocence and peace. Indeed flowers signify all that is good, pious, and noble. By culturing flowers we celebrate one of the greatest heritages that nature has bequeathed us with, and acknowledge our gratitude to nature.

Let us come to the more prosaic but scientific aspects of flowers. In botany a flower means the part of a seed plant that contains the organs (ovary and stamens) concerned with reproduction. These reproductive structures are usually associated with the sterile outer structures (sepals and petals) that protect them and attract insects and other birds. In some seed plants, such as magnolias and camellias, the flowers are borne singly, but most have flowers arranged in clusters of varying structure, called inflorescences. Quite often, the name flower is commonly given to what actually are inflorescences, such as clover and dandelion, as it is given to true individual flowers, such as the rose.

Flowers vary enormously in size from the almost microscopic to giants a meter or more in diameter. The world's smallest flowers (which feature in Guiness Book of World Records) - that of duckweed (*Wolffia*) a minute, floating freshwater weed - are so tiny that we require a dissecting microscope to enable us to see them clearly. On the other hand, flowers of *Rafflesia*, a root parasite lacking leaves, may weigh as much as 7 kg or more. In general solitary flowers are larger than those occurring in clusters.

We always associate flowers with fragrance. Indeed many flowers do prossess enchanting fragrance but there are also flowers which give off repulsive odours - such as the lily *Rafflesia* mentioned above which could smell like rotting fish, and is aptly called 'stinking corpse lily'. Of course if we steer clear of the exceptions such as *Rafflesia*, we do get fragrance from most of the flowers. They range from strong to subtle, evocative to distractive, sweet to poignant. Often the sheer visual beauty of a flower is excelled only by the enhantment of its fragrance. Such flowers, as rose, can launch a thousand poems in the hearts of even the most prosaic.

What is the source of the captivating perfumes contained in flowers? In one investigation researchers used large syringes to collect air samples from orchids and analysed the chemical constituents of the fregrance. They discovered that nothing else but the basic constituents of such common substances as vanilla, clove oil, wintergreen, and menthol are present in different scents in different combinations. A piece of paper, saturated with the proper combination of chemicals and hung on a tree, attracted the bee pollinators quite as effectively as did the flower itself. These experiments prove the rather upsetting theory that behind all the great human emotions, behind all the great human emotions, behind all the intricacies of mind and thought, there seems to be nothing more divine than a set of chemicals and a sequence of chemical reactions! As soon as we can identify these chemicals, we can manipulate the human reactions as easily as we do a chemical reactor in a laboratory!

FLOWER ARRANGEMENTS

Garlands, bouquets, wreaths and other arrangements aim

to enhance the already formidable appeal of flowers. This reminds us the Japanese art *ikebana*, because all other traditions of flower arrangements seem to have been inspired by this 1400 year-old art. It is said that the Chinese were the first to dabble in flower arrangements but it were the Japanese who developed it into a formal art form, establishing rules, masters, and schools devoted to *ikebana*. Of course modern concepts have helped in revolutionising this traditional art. After the second World War *zenei-bana*, an avant-garde trend, has spread with great popularity. Departing from set formulae, the modernists have cut themselves loose from the shackles of reality that had held the traditionists captive. Pure forms and colours are used according to their own laws by artists exercising complete liberty. Organic and inorganic motifs of all manner and description have been used, as well as abstract sculpture, mobiles, and reliefs. The more avant-garde desighs even employ scrap metal, iron wheels, engine mufflers, rope, cork, and other absolutely untraditional materials.

In terms of size also there have been attempts to scale new heights, sometimes literally. An example is a 36 feet 10 inch high boquet made from over 9000 flowers at Annecy, France, in 1986. It took 36 people one-and-a-half months to put this boquet together, breaking all previous records of size. The largest arrangement of a single variety of flower was made by a floral designer from Amstardam with assistance from 15 persons, in 1986. In consisted of 35000 Zurella roses and had an area of 45000 cubic feet - the size of a largish Kalyanamandapam!

Well, there is no limit to human ingenuity - even if it may all be due to nothing else but a few chemicals and their reactions!

4

WETLANDS : TREASURE TROVES *FACING EXTINCTION*

Life *began* in water - researches on evolutionary chemistry during the last few decades have proved this more or less conclusively. And life is *sustained* by water - no research is needed to prove it. We won't last, neither will this world, if there to be no water.

Helmets, towns, cities ... entire civilizations have grown where water was available. Water has therefore been the focus, the pivot around which human destinies have taken shape.

And water has been reckoned as life-giver, sometimes even worshipped, by most of the civilizations. In our own trinity water features as one of the three pillars of existence.

If water is sacred, if water is precious, then wetlands which *hold* water must be equally sacred, must be as zealously guarded as one would guard treasure-troves.

But strangely - and sadly - we have treated wetlands with indifference rather than care, cruelty rather than affection,

contempt rather than respect. More 'developed', and 'advanced' an area, more pathetic is the condition of its wetlands. In cities, which are the epitoms of 'development' wetlands have either been dead and buried or tottering on the brink. Wherever urbanization reaches the shores of a wetland the countdown for the wetland's demise begins. Indeed land sharks do everything possible to hasten the death of a wetland so that they could grab the land that becomes available, if possible for putting up skyscrapers.

Such mishandling of nature's bounties can't go on without backlashes. It might take time for the backlashes to become strong enough to sting but when the sting begins it keeps on increasing with frightening severity. And all those cities and towns, all those people, who had harmed their wetlands are beginning to pay the price in terms of shortage of water, flooding (when it rains), pollution, and a ravaged microclimate. More punishment is in the offing!

India is one of the most-blessed countries of the world in terms of rainfall. We receive more rainfall *per unit area* than majority of the world's countries. And inspite of our enormously large population we still receive more *per-capita* rainfall than such developed countries as Germany and Japan.

Yet we are perpetually short of water (when not drowning in it during floods). We have managed to create scarcity of water even in a place like Chirrapunji which was the world's wettest place not long ago!

Obviously there is something seriously wrong in the manner in which we have been dealing with our wetlands. Where we have not killed them we have maimed them; we have disturbed them when they were at peace and we have ignored them when they needed our care. I am proud of the fact that I am an alumni of IIT Bombay but not so proud of the fact that Powai Lake is dying right under the nose of an institution which boasts of one of the biggest centres of environmental science & engineering in India! Till early 1970s Powai was an exquisitely beautiful lake, with its emarald blue waters which came upto the Marol-Powai road and the lake-side hostels of IIT, enhancing the charm of everything that existed there. Then slowly the defiling of the lake began. Continuous dumping of sewage and garbage by the populace skirting the lake took its toll. Aquatic weeds moved in, *ipomea* struck roots, fishes dwindled. At one point during 1973 a banner was placed on the IIT side of the lake shore warning the students not to swim in the lake because the water had become too polluted for comfort (and safety). After passing out from IIT in 1975 I used to make my yearly pilgrimage to my *alma mater* and had the mortification of seeing the lake age at a frightening speed with every passing year. Today *ipomea* has so badly defaced Powai that the lake is a mere shadow of its past glory. Water is being pushed farther and farther away from IIT as *ipomea* thickens in density and conquers new territory from the lake with every passing moment. I fear the day is not far when the same type of ugly skyscrapers which have come up on the western side of the Marol-Powai road would be seen on the eastern side of the road as well - a location where the water of Powai once played with the fishes and the breeze.

When a lake which is flanked by two premier institutes of learning in India - IIT Bombay and NITIE - can face a fate as sad as Powai, one can well imagine the lot of other lakes which are situated in humbler surroundings! This, once again, underscores the seriousness of the threats wetlands are facing, and the urgency to do something about it.

Reviving the endangered wetlands is not an easy task because the problems with which they are beseiged are extremely diverse and complex as we would see from the studies embodied in these volumes. But if a determined and sustained effort is made this revival is not impossible.

And we better put in this effort, and now, if we wish to survive!

5

URBAN FORESTS

TREES AND THE MAN

Trees have been esthetically, commercially, and spiritually important to mankind since the earliest civilization. The Indians, Egyptians, Phoenicians, Persians, Greeks, Chinese and Romans - all held trees in high esteem and in certain situations even worshipped them. Vedic literature is replete with loving and respectful references to trees. They were recognized as entities which possess sensibilities of hearing, seeing, smelling, tasting, and touching; of feeling sorrow and joy in a manner human beings feel :

सुख दु:ख्योश्व ग्रहणच्छित्रस्य च विरोहणात् ।

जीवं पश्यामि व्रक्षाणां अचैतन्यं न विद्यते ।।

- Mahabharata, Shanti Ch.153. Trees were even comparted to sons :

वृक्षदं पुत्रूवद् व्रक्षास्तारयन्ति परत्र च ।

भतस्मात्रडाग सद्वृक्षा रोप्या: श्रेयोडर्थिना सदा।

पुत्रतव् परिपाल्याश्च पुत्रास्ते धर्मत: स्मृता: ।।

- Mahabharata, Anusasana, 58, 30-31.

In *Danakhanda* of *Caturvargacintamani* Hemadri observes that a person who is unable to give even a pot of water in alms, should yet water the *Asvattha* regularly - that would shake off evils and beget good progency :

उदकुम्भप्रदानेऽ पि हयशक्तो यः पुमान् भवेत।

तेनाश्वत्थनरोर्मूलं सेच्यं नित्यं जितात्मना।

- *Danakhanda*, *Ch 13.*

In all ancient civilizations mentioned above, trees were used for their material as well as esthetic benefits. Formal gardens and sacred groves were developed to enhance temple settings. Trees were even used to complement statues and provide a landscape for buildings. Along with these uses, knowledge of tree care and tree breeding was developed. The art of transplanting of trees was known as early as 1500 B.C. in Egypt.

This knowledge continued to develop as civilizations advanced. Botanical gardens began to evolve during the Middle Ages with particular emphasis on plants with medicinal properties. As the Renaissance period evolved, man embarked upon new adventures in scientific achievement and trade. As part of this increased trade and travel, plants from one country were introduced to other country. Tree species were exhibited in private gardens and collections and led to the establishment of large botanic gardens in many countries.

Then, from the middle of this century, the global attention because so focussed on industralization, development, and urbanization, that our consciousness of the importance of forests in particular and environment in general became dim. Millions of square kilometers of natural forests were cleared for agriculture, industry, and urban neighbourhoods across the world. As of now the rate of urbanization is so high that villages are turning into towns, towns into cities, and cities into mega cities. In this situation urban forests can helps us cushion the impact of the loss of natural forests.

URBAN FORESTS

Everyone is deeply concerned, and rightly so, about loss of natural forests and the need to reverse this trend. But in areas from where natural forests have been irreversibly lost - such as urban areas - one can develop 'urban forests'. As the extent of urbanization increases rapidly so does the need to pay attention to developing urban forests.

Urban forests can be entirely manmade or partly natural and partly man-made. They may include large areas of native woodlands both within and adjacent to cities and towns. There are instances in countries less populous than India where native forests have been urbanized as cities have grown into them. In these situations, the original natural forests have been so altered by the supplemental planting of introduced species that they have taken on the characteristics of manmade forests.

FACTORS EFFECTING COMPOSITION OF URBAN FORESTS

The composition of an urban forest can vary greatly according to the economic area within a particular city in which the forest comes up. In areas of low resident income, the urban forest is either composed of declining older trees remaining from times of greater neighborhood prosperity, or of trees that yield good revenue. The emphesis shifts on ornamental value and diversity in more effluent localities.

During the time of the Roman Empire, the Renaissance, or even later nineteenth century cities, those who can afford it, preferd to live on the fringe of the city absorbed by nature - although sometimes the trees were made to conform to man's idea of a more appropriate form by pollarding or topiary work. At present if one goes to cities in developed countries and seeks out the most prestigious residential areas - one will most probably find that most are in wooded sections. Names also give an indication : Forest Hills, Woodland, Lake Forest, Vrindavan, Alkapuri and a host of others that explicity or implicitly equate trees with a high quality living environment.

IMPORTANCE OF URBAN FORESTS

Urban forests are important to the city dweller in many ways. Their role in maintaining biodiversity is much smaller than of natural forests but in several other ways urban forests provide the same benefits are natural forests do. Their trees provide shade and beauty, improve microclimate, and yield materials of value. In most instances these benefits are taken for granted. Indeed, the urban dweller may not even be aware of many of these benefits, or even relate to them. The benefits can be grouped under the following four broad categories :

1. Climate amelioration
2. Engineering uses
3. Architectural uses
4. Aesthetic uses

CLIMATE AMELIORATION

Let us look at how urban forests improve microclimate. The major elements of climate that affect us are solar radiation, air temperature, air movement, and humidity. We have comfort zones associated with the interactions of these four elements. It can be too hot, too cold, or just right. The comfort zone of 'just right' varies according to individual, sex, age, and the particular climate to which one is adjusted.

Trees, shrubs, and grass ameliorate air temperatures in urban environments by controlling solar radiation. Tree leaves intercept, reflect, absorb, and transmit solar radiation. Their effectiveness depends on the density of species foliage, leaf shape, and branching patterns. Deciduous trees - or trees who shed their leaves once or more in an year - are very effective in heat control in urban settings in temperate regions. During the summer they intercept solar radiation and cause lowering of ambiant temperatures. In the winter the loss of their leaves results in the pleasant warming effects of increased solar radiation.

Trees and other vegetation also aid in ameliorating summer air temperatures through evapotranspiration. For this reason trees have been called 'nature's air conditioners'. A single isolated tree may transpire approximately 88 gallons (400 liters) of water per day (provided that sufficient soil moisture is available). The effect of this has been compared to that of five average room air conditioners, each with a capacity of 2.5 tons, running 20 hours a day.

WIND PROTECTION AND AIR MOVEMENT

Air movement, or wind, also affects human comfort. The effect may be either positive or negative depending largely on the presence or absence of urban vegetation.

During summer, air movement has relatively little effect on air temperature unless the wind is part of a cool front. Instead, it simply causes a cooling sensation because of convective heat loss and evaporative cooling. Because trees screen sunlight and transpire moisture, the area below the forest canopy can be as much as 14°C cooler on a still summer day than an open area.

Trees reduce wind velocity and create sheltered zones both leeward and windward.

In many parts of the world, trees are used as effective wind controllers. This is particularly true in the Great Plains and other relatively treeless areas where severe winds cause discomfort, irritability, and occasional loss of life and property. In these regions, a belt of trees to slow the wind is a welcome relief. Beach erosion due to winds can also be significantly controlled by efforesting the beeches.

ENGINEERING USES

In recent years, highly specialized uses for plants in solving environmental engineering problems have been developed. In it are involved not only landscape esthetics but soil erosion control, air pollution, noise abatement, wastewater management, traffic control, and the reduction of glare and reflection. One may cite the following plant characteristics and their effects that help to solve environmental engineering problems :

i) Fleshy leaves that reduce noise.

ii) Branches that move and vibrate to absorb and mask noise.

iii) Pubescence on the leaves to entrap and hold dust particles.

iv) Stomata in the leaves to exchange gases.

v) Blossoms and foliage that provide pleasant smells to mask malodors.

vi) Leaves and branches to cushion the effect of heavy winds.

vii) Leaves and branches to slow the eroding effects of rainfall.

viii) Spreading roots to hold soil against erosion.

ix) Dense foliage to block light.

x) Light foliage to filter light.

xi) Spiny branches to deter human and other animal encroachment.

ARCHITECTURAL USES

In building design, materials such as wood, masonry, steel, or concrete are used architecturally as well as structurally. Each tree species has its own characteristic form, color, texture, and size. Plants can vary in their use potential as they grow or as the seasons change. Their proper use will vary with the designer and user. Trees, when used in a group, can form canopies or walls of varying texture, height, and density. Some functions can be performed by one tree and others may require many trees. Because they are alive and growing, trees and shrubs are dynamic with regard to their functionality in architectural design.

Since trees and shrubs have architectural potential, they can be used individually or collectively as architectural elements performing the following functions : space articulation (defining space); screening; privacy control; and progressive realization or enticement.

AESTHETIC USES

Trees and shrubs provide their own inherent beauty in all settings. They are esthetic elements in our surroundings. They can be beautiful simply because of the lines, forms, colors, and textures they project. Trees and shrubs enframe views, soften architectural lines, enhance and complement architectural elements, unify divergent elements, and introduce a naturalness to otherwise stark settings. They also produce unique ambiances through reflection from glass and water surfaces and can produce beautiful shadow patterns. Trees are also dynamic, giving different appearances in the changing seasons and throughout their span of life. They also provide mellow and pleasant sounds such as the rustling of leaves and the whistle of wind through a canopy. We are attracted to plants because of many of these characteristics. In addition, plants are useful for the berries, nuts, and shelter they provide to birds and animals.

OTHER USES

Although a number of benefits of the urban forest have been mentioned, there are others. Firstly, there are economic products such as fuelwood, fruits, nuts, christmas trees, pulpwood, and sawlags. Less tangible, but perhaps no less obvious, are the values of places for childern to play, for people to jog or to walk and contemplate nature (or their own problems), for groups to stroll, or just for one to be alone. Trees are also used as indicators of historic events, as memorials, and as substitutes for the natural environment in inner cities - even on rooftops and balconies. Trees also evoke memories of other times, places, and feelings because of the view they present, or a familiar sound, smell, or touch. Who indeed can walk on a leaf cluttered sidewalk on a rainy autumn day and not become nostalgic by the sight and smell of damp leaves and the patter of raindrops?

Studies on the perceived psychological benefits of urban forests and related green spaces indicate that these area provide a variety of subjectively evaluated benefits, including :

i) developing, applying, and testing skills and abilities for a better sense of self-worth;

ii) exercising to stay physically fit;

iii) resting, both physically and mentally;

iv) associating with close friends and other users to develop new friendships and a better sense of place;

v) gaining social recognition to enhance self-esteem;

vi) enhancing a feeling of family kinship or solidarity;

vii) teaching and leading others, especially to help direct the growth, learning and development of one's children;

viii) reflecting on personal and social values;

ix) feeling free, independent, and more in control than is possible in a more structured home and work environment;

x) growing spiritually;

xi) applying and developing creative abilities;

xii) learning more about nature, especially natural processes, man's dependence on them, and how to live in greater harmony with nature;

xiii) exploring and being stimulated, especially as a means of coping with boring, undemanding jobs; to satisfy curiosity and the need for exploration.

Trees have been called 'lungs of the earth'. It will be no exaggaration if we say that trees are not merely the lungs but also the arms, legs, and the crowning glory of the earth!

6

THE WEEDS OF DESPAIR AND HOPE

When rains come, so do the aquatic weeds. No sooner do the drought-hit tanks, ponds and canals get their fill of water, the weeds begin to move in. And we often see carpets of water hyacinth, salvinia, or water lily rather than the face of water in the water bodies. The shoots of weedy grasses *Scirpus* and *Cyperus* which stand on the fringes of the weed mats like body-guards, also come on, presenting a formidable sight for the fishermen, farmers, and boatmen everywhere in India.

It is estimated that nearly half of the total inland water-surface in India gets under aquatic weeds during the days that immediately follow the onset of monsoon. The weeds cause incalculable harm to the water resources - in terms of quantity as well as quality. They reduce the storage capacity of the tanks, ponds and canals by literally crowding out the water. They also hasten the loss of stored water by using it for their own growth as well as causing excessive water loss through evaporation via the moist surface of their leaves and other

tissues - a process known as 'evapotranspiration'. The weed mats suppres the normal waves and currents that should be arising in any water body and cause the water to stagnate. Their blankets cut off sunlight from reaching water. These happenings along with the rot caused by the decay of old weed plants play havoc with the water quality. A well-aerated healthy water environment starts turning into a cesspool - inviting mosquitoes and snails and discouraging fish growth.

The two aquatic weeds which, between themselves, have garnered the bulk of water surfaces in the world are water hyacinth (*Eichhornia crassipes* Mart Solms) and salvinia (*Salvinia molesta* Mitchell). Water hyacinth is believed to be the most widespread and problematic aquatic weed of the world but in many countries it appears to be losing ground to salvinia. Indeed salvinia has moved even into the Guiness Book of World Records as 'the world's worst weed'.

Between the two, salvinia looks by far the more innocuous and fragile. While the robust looking and wayward stems and leaves of water hyacinth rise well above the water surface, sometimes upto as high as four feet, salvinia keeps itself much to the water level. While water hyacinth is uneven and coarse in its growth salvinia spreads evenly and harmoniously like a carpet laid out on water and earning for itself the sobriquet of 'velvet-weed'.

A lay onlooker may never believe that salvinia can elbow out water hyacinth. Yet, this is exactly what salvinia has done in Kerala, Goa, and in many other parts of the world. This may appear incredible when we recall that water hyacinth is so hardy and combative that it has been virtually fighting and winning against the army of United States of America! Unbelievable, yet true, Water hyacinth was first brought to North America as a souvenir by Japanese deligates attending a trade fair in 1884. While presenting the exotic plant to their American friends the Japanese politely mentioned that the violet flowers of the aquatic may add to the beauty of the American lakes and ponds, especially in the warmer Southern United States.

Indeed, and so the Americans found to their delight, water hyacinth produced enchanting blooms of violet flowers. But the delight soon began to turn to apprehension, then dismay, and then despair when the yankees found that the plant does not stop at producing its violets but keeps replicating itself at a monsterous pace. Instead of decorating the lakes and ponds, hyacinth was *overwhelming* them by crowding out all other vegetation and enveloping the water out of sight.

This wont do, the Americans thought, and sought to clear off the plant from wherever they had introduced it. Then, to their horror, they found that the plant was in no mood to leave! With great effort they would clear a pond of water hyacinth and bang will the plant come back like Baital's ghost! Later researches have revealed that even small pieces of water hyacinth leaves or an odd spore can trigger off the replication processs - therefore it is well-nigh impossible to mechanically clear any water body entirely from water hyacinth.

So, water hyacinth continued to spread explosively. The matters reached a head in 1900 when a massive 160 km long water hyacinth colony was discovered in the process of further expansion. In desperation army was called in to remove the weed. The army, however, lost the battle and hyacinth is still there, making Americans sink over 11 million dollars annually *just to keep the weed in check in just three of its fifty* States.

Of course as soon as the weeds spread their tentacles the man rose to fight them but so far it has been an uneven battle and man has been at the losing end most of the time. Paradoxically it is man who has helped weeds to colonise new countries by introducing them as a research or aquarium plant. Just as water hyacinth was taken across the Pacific from Japan to America by human beings, salvinia was taken from Australia to Papua New Guinea in early seventies by amateur aquaculturists for decorating their aquaria. As the aquaria were covered up by salvinia in no time the aquaculturists, still unsuspecting, threw away the excess plants. These discards got into canals, rivers and lakes and by early 1977 salvinia was everywhere on the water fronts of Papua New Guinea. The thick stable mats of the weed became virtually impenetrable for native dugout canoes thus denying many island villages access to their river gardens, fishing and hunting areas and local markets. It became difficult for children to go to school and sago palms to be collected. Life was completely disrupted and rendered miserable.

The weapons that man has tried on the weeds make a long and varied list. They include pesticides of all manners and description, mechanical removal methods, and bioagents. Pesticides do kill the weeds but seldom completely. Worst still they are very costly and themselves pollute the water, thus constituting a major health hazard. Mechanical removal methods are as ineffective because reinfestation occurs with a vengeance. Also, mechanical removal puts forth two big questions

- how to cover up the heavy costs involved and what to do with the harvested weeds. Bioagents appear safer and cheaper and tremendous efforts have been made to find the animal which can devour the weeds faster then the later can grow. This search has led man to commission fishes, ducks, seacow, snails, beetle, grasshoppers etc. Unfortunately the common story of each of these animals is - they came, they saw, they were conquered. In 1981 a report came from Australia, published in 'Nature', that the beetle *Crytobagous singularis* has knocked out salvinia mats of Lake Monderra, However the beetle does not seem to have been as successful when but in action elsewhere. The biggest danger with bioagents is that under favourable conditions they themselves can multiply and grow explosively, becoming a 'weed' themselves.

As it is becoming more and more clear that weeds will take some beating, man is trying to find ways and means of welcoming the weeds rather than fighting them. Attempts have been made to utilise weeds as compost, as livestock feed, as source of chemicals and as paper pulp. Composting takes time and effort and poses a danger of causing perpetual reinfestation. Livestock do not relish salvinia and water hyacinth unless they have nothing else to eat; besides the nutritive value of the weeds is not adequate. Pulping causes environmental pollution problem of its own. The basic problem in finding methods of utilisation of aquatic weeds is their high water content which is in the range of 92-96%. Whatever they have - be it useful chemicals, nutrients, pulp - is in a very dilute form making recovery methods eneconomical. After exploring myriad utilisation options this author and coworker Dr P.C.Nipaney zeroed in on tow possibilities - generation of biogas and waste-water treatment - which are technically as well as economically feasible. In fact the two are complementry too.

Studies encompassing over 12 years, and reported in a recent book ('World's Worst Weed - Impact and Utilisation', International Book Distributions, Dehradum 1993) reveal that with specially designed digesters salvinia and water hyacinth can be used as source of biogas both singly and as weed-cowdung blends. Digester tests show that a 50.50 mixture of salvinia and cowdung as feed gives the best results yielding biogas which is over 20% higher in quantity compared to biogas obtainable from cowdung-fed conventional digesters. The calorific value of the gas is as high as of biogas from cowdung.

Both weeds are excellent bioagents for treating sewage and other biodegradable wastewaters. Indeen water hyacinth based wastewater treatment plants have now become fairly common in USA and Japan.

Calculations based on the average growth rate, spread of salvinia, and the convertibility of the weed intoi biogas reveal that in kerala alone salvinia - fed digesters can produce biogas worth Rs 13,000 lakhs per year. Judging from the fact that several countries including Sri Lanka, Australia, Papua New Guinea, Indonesia, Thailand, Burma, Kenya, Rhodesia and Zaire are confronting the salvinia problem, and that water hyacinth is present in all the rest of tropical and sub-tropical regions, the applicability of these utilisation options is likely to be sizeable.

7

THE MOST IMPORTANT ANIMAL ON EARTH ?

Which is the most important animal on earth?

The horse?

No, eventhough the horse might well have shaped the history of mankind. After all, each and every of the great medieval wars were fought on horseback. It was the gallantry of the British horses as much as of the British soldiers that decided the battle of the Waterloo - a battle that had such far-reading impact on the colonial history, including that of India. Even today, wast fortunes are gained or lost daily on the basis of the horse's abilities in the race courses all the over the world.

The holy cow? The industrious bull? The exquisite peacock, the sonorous cuckoo? They are important no doubt, very important too - as workforces or aesthetic delights. But none is *the most important.*

Not also the king of the jungle, The Lion - who sits right on top of the food pyramid. Nor the fish which provides to us the most easily digestible and versatile source of animal protein. And, most certainly, not the humans because inspite of all their progress in the fields of science and technology, inspite of the control they have gained on nature, human beings still remain the only animals on earth who kill their fellow beings not out of necessity but out of jealousy and greed.

The most important animal on earth is so small and slimy that we, humans, refer to it derogatrily as synonyms for meekness and ugliness. It is an animal which can't even sting like insects do, and if we decide to stalk it, it wont even be able to run away from us to save its life.

And yet it is this animal who tills, 'purifies' and fertilizers enormous masses of soil for us, who rejuvinates our agricultural lands after incessant use has sapped their productivity, and whose efforts ensure that these fields do not become barren and sterile - that animal has no limbs, no teeth, no backbone. It is but a tiny loathsome creature, often used as a symbol for despicablity and spinelessness. This creature is the earthworm which works so silently and unobstrusively at our feet that we have taken very long to realise its real worth and bestow over it the attention it deserves. According to Hartenstein, one of the foremost earthworm technologists working in USA, 'The earthworm is the most important animal on earth!' Yes, not one of the most important but the most important.

While we are scaling the heavens with the miracles of modern technology - the satellites and the spacelabs - we have also been compelled to look right under our feet for the super 'tractor' which is capable of conserving and activating our earth as naturally and as safely as no other device is able to do. That wonderful machine happens to be the old worm and scientists the world over are now busy harnessing the natural abilities of the earthworm in reviving the agricultural fields and cleansing the ever-increasing loads of waste that are piled over the earth everyday.

Earthworms devour soil. As the soil passes through the worm's guts it gets naturally processed by the enzymes and micro-organisms present in the guts. The resulting castings, which may loosely be called 'digested soil', are richer in utilisable nutrients, enzymes and miroflora. It has been shown that earthworm castings contain up to twice as much available nitrogen, seven times as much available phosphorous, and eleven times as much available potassium as the surrounding soil. Bacterial growth is also greatly stimulated which helps in the creation of soil - supporting humus. Most important, in this process the ecessive acidity or alkanity of the soil gets reduced and is brought close to the pH of fertile soil. The constant burrowing of earthworms open up the fertile soil to let it 'breathe' air and absorb water much deeper than it otherwise would. In loose soils earthworm push their anterior portions in crevices and then bore in slowly, forcing away the obstacles. When confronted with compact and tough soil layers, the earthworms just eat their way through. The works can burrow very deep into the soil, some species can go to depths which are much beyond the reach of the plant roots. From such depths the worms buring up life sustaining minerals for use of the plants. No wonder then, that introduction of a large number of earthworms into agricultural land has been shown to double the yield of wheat, quardruple the yield of grasses and multiply the clover yields ten times. When introduced into a plot otherwise unsuitable for growth of herbal plants, the earthworms caused a 10-fold enhancement in the yield. In Netherlands it was found that garden trees with earthowrms placed around them grow heavier root systems than the nearby trees, while Soviet researchers report that in forest soil earthworms increase the growth of two - year oak seedlings by 26% and of green ash seedlings by 37%. Direct correlation has been reported between the number of earthowrms and growth of barley in a series of controlled experiments.

WASTE CLEANERS

From the recognition of the earthworms' contribution as soil conservers to their use in the treatment of organic waste is but a small step. However, mankind has taken thousands of years to take this step. Canada was the first country to install a waste consumption facility based on 'earthworm power' in 1970. During a recent tour of Canada this author visited the facility which is now processing over 80 tons of refuse per week. The refuse consists of all kinds of organic waste, including household garbage, city refuse, sewage sludge, waste from food and fibre industries and different forms of paper waste ranging from old telephone directories to outdated office memons. With the success of this facility, several more similar waste treatment facilities are now planned to be installed throughout Canada. In Japan too, during an earlier visit, this author had the opportunity to see earthworm-based pollution abatement facilities. The difference between Canadian and Japanese facilities was that the later treated specific types of industrial waste up to 10 tons per day. Earthworms do nearly all the work in these treatment facilities. The continuously turn and aerate refuse, break it down, and process it into rich manure than can be safely and profitably recycled back into the earth.

The nitrogen present in the world's sewage and fubbish is of the order of 5 million tonnes which is nearly equal to the nitrogenous fertilizers that the world's chemical industries are capable of producing. Earthworms have the potential of converting this organic waste into fertilizers and thereby doubling the world's fertilizer production. In addition they can compensate the loss of utilisable nutrients in an agricultural field by their acts of tilling and 'digesting' of soil through which they convert the 'junk' present in the soils to utilisable nutrients. During November-December 1985 the author witnessed a demonstration process in Southern California, USA, where 10,000 kilograms of rubbish ranging from old magazines and diaries to straw and grass clippings was converted into fertilizers by an army of earthworms in a matter of 40 days. The entire process is virtually free of operation or maintenance costs because the worms do all the work and 'provide' all the energy needed. They also contribute to the disinfection of sludges by reducing the survival of pathogenic bacteria. It appears that the microflora within the earthworm's gut are able to outcompete the pathogenic organisms. The success of the California demonstration process has spurred the Americans to go for vermicomposting' in a big way and now over a lakh earthworm ranches are in operation where selected species of earthworms, especially the 'redworms' (Eisenia foetida) and 'nightcrowlers' (Lumbricus terrestris) are raised and sold. Philippines has also taken to earthworm technology in a big way.

FOR FOOD

Earthworm have been the pet food for poultry and a source of delight to poultry farmers who learn that the higher amounts of certain amino acids in the earthworms can reduce the cost of feed grain for poultry by more than half. Nutritionists rank earthworm very high as a protein-rich yet easily digestible food item. But will humans ever include earthworms on their menu? It is food for thought. We, as a species, are notoriously fussy about the species we allow on the dinning table. Of the tens of thousands of species of plants and animals available, we eat only a few dozens - the rest we tend to reject, many times inexplicably. Earthworm is likely to be one of the family of animals we can benefit from eating but before we begin doing that, we may have to cross a major psychological barrier. Attempts have been made in the Philippines and Taiwan to develop recipes based on earthworms; culinary competitions have been held and a few hotels have actually begun to dish out earthworms but the little animal is far from catching the gourmet's fancy.

EARTHWORM TECHNOLOGY IN INDIA

In India domestic animal - buffaloes, cattle, hogs and chicken-produce over 400 million tonnes of manure; only a small percentage of which is used for biogas, fuel and fertilizers. The collection, treatment and disposal of household garbage puts a major burden on all over resources, and the failure to do so effectively in most of the cities and towns in India is the prime cause of filthy conditions and health hazards prevailing there. Any technology which requires little establishment costs and which converts waste material into manure usable near the site of its production, is highly appropriate especially to the Indian situation. Earthworm based waste treatment is such a technology.

We were rather slow, compared to the westerners and the South-East Asian countries, in harnessing earthworm power but over the last decade the Indian initiative has steadily grown in momentum. R&D inputs have come from senior earthworm biologists such as Prof Madhap K. Dash, Dr Abdul Ismail, and Dr Radhakale. Entrepreneurs such as Yashpal Suhag of New Delhi, have demonstrated that earthworm breeding can be as commercially rewarding as it is environmentally beneficial. There are also pathbreaking contributions such as of Dr Uday Bhavalkar and Mrs Vidula Bhavalkar who have put their all - their education, training, and monetary resources - in earthworm R&D. Dr Bhawalkar is the first engineer in India to have taken a doctorate (from IIT Bombay) in earthworm biotechnology. He and Mrs Vidula have established the first ever R&D institute in India entirely focused on earthworms - BERI (Bhawalkar Earthworm Research Institute). The biofertiliser they produce from poultry waste using earthworms have been named by the 'biogold', and the world has quickly realized that it is indeed as golden as the Bhavalkars claim it to be. As a result multinationals are now approaching BERI for know-how to utilize their wastes and turn it into well-earned money.

8

TROPICAL RAIN FORESTS

Of the various types of natural forests, the richest, most spactacular, and most fragile are tropical rain forests.

India is one of the nations of the world bestowed with large tracts of tropical rain forests. What is so special about them?

Tropical rain forests have been universally accepted as cradles of evolution. It is now believed that about half of the species known to mankind occur in tropical rain forests although such forests occupy only about seven percent of the world's land area. One of the reasons for the great species diversity is that co-evolution of plants and animals is facilitated in the rain forests, resulting in close mutually sustaining relationships.

CHARACTERISTICS OF TROPICAL RAIN FORESTS

The soil in the tropical rain forest is lean in nutrients, perhaps due to the very high productivity of the rain forest. Several species that thrive in rain forests are found to have adoted to the soil infertility. Such adoptations are rarer in temparate forests.

The follwing features characterise a tropical rain forest.

a) It is like a gaint ion-exchange column that extracts nutrients from water passing through the ecosystem.

b) The forest develops a dense root mat on the soil surface. The rootlets do not have a well developed geotropic response, and they soon grow over and cover any litter that falls on to the root mat. This root mat has a very high nutrient retention capacity. In some tropical rain forests up to 60% of the biomass is in the form of roots.

c) Direct nutrient cycling occurs from the litter to the roots *via* fungi *Mycorrhizae.* The fungal hyphae rapidly invade freshly fallen litter, absorbing inorganic nutrients before they can be leached into the soil and conducting them directly to the roots.

d) Nutrients are conserved within the trees by efficient internal cycling before leaf shedding.

e) Reductin of herbivory occurs through the accumulation of toxic secondary metabolic chemicals in leaves and roots. Annual loss of leave area to herbivore consumptions is only 1-2%.

f) The multilayered structre of the rain forest conopy, together with the presence of many epiphyllous organism such as mosses, algae, lichnes and bacteria promote uptake of nutrient from precipitation.

g) The high concentration of secondary plant chemicals in the litter of plants impedes decomposition other than *Mycorrhizae* fungi, it also restricts the acitivity of bacteria. This ensures that virtually all the nutrients released through decomposition return to the plant via *Mycorrhizae* by the direct nutrient cycle.

IMPORTANCE OF TROPICAL RAIN FOREST

The Tropical rain forests have the potentiality to change and alter the global climate by heat balance, surface roughness, the hydrological cycle and the production of various gases, notably CO_2.

Tropical rain forests contain many wild fruit trees, some of which are the ancestors and relatives of cultivated species and many are species of medical importance. Furthermore, when fossil fuels are eventually depleted, mankind is likely to make greater use of plants as source of complex organic molecules, aften as raw materials for manipulation. For example, the trunk of *Copaifera lansclosfi,* a leguminous tree of the Amazon tropical rain forest produces an inflammable oil at a rate of 20 litres per 6 months, which is tapped and used locally instead of kerosene. Indeed, forests are a biochemical storehouse, scarcely yet exploited. The provide innumerable species that are used locally in traditional medecine. The seeds of the Australian rain forest legume, the Moretan Bay chestnut, *Castanospermum australe,* have recently been discovered to contain a drug that might help combat AIDS.

ANTHROPOGENIC IMPACT ON TROPICAL RAIN FORESTS

Great concern has been expressed by many scientists about the exploitation of tropical rain forest for timber and the clearing of forest for shifting cultivation or permanent agriculture. The native agricultural traditions in the tropics involve cutting down a small patch of forest, burning the slash and planting crops in the ashes that enrich the soil with nutrients. This produces one or two good food crops, but the yield declines rapidly. The nutrients of the ashes are soon leached away or are removed in the harvested food plants, and the pH of the soil declines towards its original acidic value. The area is than abandoned.

If the interval between successive cut and-burn is not too short, this type of land use can continue indefinitely without degrading the productivity of the site. However because of the population growth, the interval between successive cut-burn-harvest events has been rapidly decreasing since world war II and in some areas the soil has became so impoverished that the original forest species can no longer grow. Agriculture is the main purpose for which rain forests are cleared. Conversion of tropical rain forest to pasture is especially widespred in the neotropic, where there is a long traditon of cattle husbandary. The Atlantic coast rain forest which once covered one million sq.km. has been 99% cleared for pastore, mainly over the past three decads.

Rain forests have also been felled and replaced by plantation tree crops, principally rubber and oil palm. Most of the rain forests are primarely exploited for timber export. Dams construction, mines exploration, wood chips for rayon (Asian mangroves wood chips are used for rayon in Japan), fuelwood and chorcoal preparation also lead to severe destruction. Tropical rain forests are very major CO_2 sinks. Loss of these forests is one of the most significant causes of increasing greeenhouse effect.

The Food and Agricultural Organization (FAO) estimates that some twelve million hectars (ha) of closed forests are being cleared per year. The rate at which rain forests are being currently destroyed is causing their virtual disappearence from some places, for example Coasta Rica, Sumatra and the Atlantic coast of Brasil.

Due to natural forest fire in early 1983, some 3 million ha of forest in Kalimantan tropical rain forest and over one million ha of Sabah rain forest were destroyed. But even vaster areas of rain forests have been deliberately destroyed by fire to create pastore on the southern fring of Amazon in Brazil estimated as 8 million ha in 1987 and 8 million ha in 1988. High intensity burns for forest clearing purpose can produce ozone concentration in excess of 100 ppb in one hour.

Only part of the forest goods are seen to be important in the contemporary world and little consideration is given to the services forest also provide and on which it is hard to put a cash valve. When man reduce rain forest to small fragments or over-exploits them, there is a loss of complexity and diversity, the web of interconnections begins to break apart. Species number of both plants and animals are reduced. Where destruction is total, extinctions probably occur because all rain forest contain local endemics of limited geographical range. Reduction within species of either the numbers of individuals or of gegrapical area is a more insidious loss of bio-diversity, sometimes evocatively called as genetic erosion because of the likely loss of ecotypes and other genetic variants. **For example, were the American rain forest to be reduced to half their original extent by the end of the century, reasoned calculation predicts the loss of 13,600 plant species and 15 per cent of the bird species.**

INDUSTRIALIZATION VERSES RAIN FOREST

Due to rapid industrialization, the important pollinating insects and ants are affected by the toxicity of arsenic, fluoride, sulfur dioxide and ozone. Many studies have proved that air contaminants reduce pollen grain retained on stigma, pollen germination and pollen tubes reaching ovules. Acid precipitation may accelerate leaching of nutrients from forest foliage and forest soil, and alter the weathering rate of forest soil minerals. Heavey metals decrease mycorrhizan fungi density, vitally involved in the nutrients cycle.

CONCLUSION

The removal or destruction of the vegetation in such nutrient deficient ecosystems as tropical rain forests will result in rapid loss of nutrient capital and can convert forests to wet deserts. We need to remain well aware that forests are naturally under continuous renewal and that if man works within the limits of the ecosystem they are a continually replanishable resource.

9

STATUS OF ENVIRONMENTAL EDUCATION IN THE INDIAN UNIVERSITIES : *A LIKELY THIRD WORLD PATTERN*

This chapter examines the present status of environmental education in India at the college/university level. Eventhough the study is confined to the Indian situation it is likely to reflect the patterns existing elsewhere, especially in the Third World.

The author finds that environmental education is besieged with two kinds of problems. Of the first kind are the problems that are associated with **all** our degree programmes as reflected by the deteriorating quality of the graduates we are producing. The second type of problems are specific to environmental education. They emanate essentially from the all-pervasive nature of environmental education which tries to draw upon, all at once, too many disparate disciplines. This causes the course programmes to spread themselves too thin, at the cost of substance and focus.

A comparison is made between environmental education and computer education - the two branches of knowledge which began evolving into independent disciplines at roughly the same time, and whose importance is continuing to increse by the day - to further delineate the problems afflicting the former. Attention is also drawn towards the confusion often displayed by environmental science graduates in their perceptions of the real-life conflicts between development and conservation.

In the concluding part the paper makes an effort to initiate debate on the possible ways of remeding the present situation.

INTRODUCTION

Till a couple of decades ago there wasn't a single multidisciplinary department of environmental studies anywhere in India. Whatever environmental research and teaching was done in our country till the mid 1970s was all confined to uni-disciplinary groups. Ecology was considered as one of the specialisations of botany. Strangely even limnology - which draws upon myriad branches of physical and natural sciences and has only a small share of botany - was practiced by experts who had their formal training in botany. Environmental research *was* being done by wild-life zoologists, geographers, and others but it was bracketed as part of the respective parent discipline.

In the field of technology, environmental teaching and research was largely the prerogative of the public health engineering departments of engineering colleges and universities. Water supply and sanitary engineering was the basic discipline where, off and on, some R&D was getting done in areas which as of now would be considered as belonging to environmental engineering.

This situation began to change dramatically in the mid 1970s. Environmental pollution had become a *nouveau vogue* in the developed countries following the publication of Meadow's Limits to Growth and the doomsday predictions of Club of Rome. This new environmental awareness soon began to envelope India as well. Reports began appearing in the media on the high levels of carbon monoxide in Chembur (Bombay), yellowing of the Taj, and DDT in human milk. Terms such as *ecosystem, holistic approach*, and *environmental degradation* began to feature with increasing frequency in the conversations of the articulate. Phrases such as *development without destruction, pollution prevention pays* and *our only environment* began to be increasingly heard in elite gatherings.

Around that time India witnessed its first - and to date the most famous - environmental movement : the **Chipko Aandolan**. This movement not merely sensitised the Indian people about environmental protection, it had enough force to catch global attention and become a part of the conservation lore the world over. Chipko Movement was followed by the Chaliar Movement. Whereas the first movement was against the plunder of forests by those perceived as the agents of the government, the second one was to protect a river from unscrupulous industrialists. In the years to come forest and water were to become the recurring themes in people's struggles against environmental degradation.

Thus, during mid-1970s, the stage was set for the 'greening' of our educational system to begin. Schools of environmental sciences were set up at JNU and Poona University. Centre for Environmental Science & Engineering came up at IIT Bombay. IISc started its Centre for Theoretical Studies headed by a Professor of Ecology. On the technological front the public health engineering courses were refurbished and remoulded into environmental engineering programmes. Several new colleges began offering ME/MTech in environmental engineering. Central Public Health Engineering Institute (CPHERI) was rechristianed National Environmental Engineering Institute (NEERI).

Even as sceptics ridiculed the new wave by calling it such names as *environmental fashion* and *environmental bandwagon*, the movement surged ahead with unabated vigour.

THE SITUATION AT PRESENT

By now we have over 200 schools/departments of environmental studies in the universities and colleges all over India. They offer degree/diploma programmes covering all aspects of environmental sciences and engineering. There are diploma/MSc/MPhil/PhD programmes in environmental sciences or environmental studies - some with emphasis on natural sciences, some on physical sciences, some on social sciences. There are ME/MTech/PhD programmes in environmental engineering offered by civil engineering departments, chemical engineering departments and, in some cases, even by mechanical and mining engineering departments. Then we have postgraduate degree programmes in environmental management and courses which lead to MSc in environmental chemistry/ environmental biology/environmental geology/environmental toxicology etc. Even bachelor's degrees in environmental sciences and environmental engineering are on offer. More and more new programmes are being started every year.

Environmental education at doctoral level is also available in a large number of autonomous R&D institutions funded by central government, state governments, and agencies such as CSIR, ICAR and ICMR. In a word, it is now possible to study practically every branch of environmental science or engineering in India upto the highest possible level.

THE INPUTS

What inputs typically go to a college or university-level department of environmental science in India? How adequate these inputs are in terms of faculty and equipment?

Till the environmental science schools/departments began producing PhDs the faculty of the nascent departments/ schools was almost entirely drawn from persons trained in conventional disciplines with *some* exposure or aptitude in environmental studies. Even today whereever new departments in this field are started they are largely manned by persons trained in conventional disciplines as the number of faculty required to teach environmental science is much larger than the number of suitable PhDs available in this field.

Indeed, it is inevitable that whenever a new discipline is initiated it has to be done so by experts trained in other disciplines. Even during the first few decades of evolution of a new discipline the inputs come from experts drawn from other, more conventional, fields. Where environmental science differs from other new multi disciplinary areas of comparable age - such as computer science, aeronautical engineering, operations research - is in its vast canvas which envelopes practically anything and everything. Not merely the world but the whole universe can be brought under the gamut of environmental studies. All known disciplines seem to come under its sweep. This 'limitless versatility' of environmental studies is a blessing as well as a curse because it bestows upon the new discipline not only a great reach but also great confusion and, in terms of academic rigour, a propensity for great dilution.

A CASE IN POINT

A case in point is Salim Ali School of Ecology established at Pondicherry University in 1987. We are chosing this School to illustrate our points for various reasons. First and foremost reason, of course, is that it belongs to our university and we do not run the risk of antagonising our colleague from other universities by citing examples of their discomfiture. The second reason is that the school belongs to one of the central universities which are traditionally believed to be better funded than state universities. The third reason is that the problems of the school, in their magnitude and complexity, provide an apt case in point. Indeed when one of the authors (AG) took over as VC of Pondicherry University, one of the most complex problems he inherited was of this school and it took tremendous efforts, as also a great deal of reorganisation, to bring a semblance of order into its functioning.

As this school came into existance a full 12 years after the first-ever such school in India, we had the benefit of hindsight. Yet the school embodies all the problems and confusions that are still deeply embedded with environmental education in India. You may recall that in the early 1970s - when it all began - there used to be endless discussions and undiminishing heat on the very meaning of the term *environment*. Everybody was sure of what it meant and yet nobody was sure of what it meant! In School of Ecology the first and the fiercest struggle began for the distinction between 'ecology' and 'environment'. The school had recruited 10 regular faculty members and two adjunct professors of whom none had a bachelor's or master's degree in ecology or environmental science. One of the regular faculty members had done her PhD with Salim Ali and another in environmental engineering. One adjunct professor had his PhD in wildlife ecology. All others had their formal education right upto doctoral level in conventional fields. With no formal label to assist them the faculty members labelled themselves into 'ecologists' and 'environmental scientists'. Simultaneously a battle royal began on which group should dominate the proceedings in the school! 'Ecologists' claimed that environmental science is a sub-discipline of ecology whereas the 'environmental scientists' demanded that ecology being 'one of the domains' of environmental studies -

with better job opportunities to boot, they claimed - the curriculum should be slanted towards 'environment' rather than 'ecology'. The 'ecologists' said that all this holds no water or else the school ought to have been named School of Environmental Sciences and not School of Ecology. The 'environmental scientists' came back with the question - if so why were we recruited at all?

In retrospect we may find all this laughable but it wasn't so funny as long as it lasted. Indeed even today the strife has not been totally resolved. Only, the protagonists having burnt off much of their passion are less stridnt now.

If a 'school of excellence' can have so much conflict over mere terms what would happen of the admission policy, curriculum, infrastructure? Expectedly all these had a tortuous birth and painful childhood. The faculty had had their training in diverse disciplines ranging from social science to environmental engineering and everyone wanted his/her share of the spoils. So the student intake was made multi-disciplinary to such an extent that even bachelor's degree holders in english literature and history found admission. The 'ecologists' strove to orient the curriculum towards field work in deep forests across high seas while 'environmental scientists' pressed for rigorous training in instrumentation and other niceties of labwork. The end-product was akin to a cross between a zebra and a camel - an awkward creature with a little of everything but little of substance. Teachers found themselves addressing students who had their graduation in widely desparate disciplines. If a statistics lecture was too easy for the students who had graduated in physics or engineering, it was too difficult for those who came after having studied history or English literature. If biology was meat and drink for bioscience graduates it was an unsufferably descriptive and tongue-twisting experience for physics or engineering graduates. The teachers therefore began to take the least common denominator. The result was the inevitable dilution of the course content.

When one of the authors (AG) joined Pondicherry Universsity in early 1991 the School of Ecology had ran for 3 1/2 years. It had already witnessed four changes of Deans. Some of the students who had joined its first batch in 1987 and were supposed to have passed out by mid 1989 were still there, still working on their master's theses! We are sure Salim Ali School of Ecology is a rather extreme example of what has gone wrong with environmental education in India. Things must be certainly better in most other schools/departments. Even the Salim Ali School of Ecology, after it was reorganised into Salim Ali School of Ecology & Environment Sciences with its faculty trifurcated into two centres and one department, has gained a great deal of stability and direction. But, we are equally sure, the situation is far less comfortable in the domain of environmental education than, say, computer science education. We have cited computer science education for making this point for two reasons. First is that environmental science and computer science are roughly of the same age as formalised disciplines. Secondly both have been high-priority, and fast-expanding areas of whose importance seem to be growing all the time.

PROBLEMS ASSOCIATED WITH ENVIRONMENTAL EDUCATION IN INDIA

The problems faced by environmental education in India come under the following two broad categories.

a) Problems which are common to the education of *all* disciplines in India;

b) Problems which are specific to environmental education.

THE COMMON PROBLEMS

There is no quantitative study available, or of which I am aware, of the *quality* of output of our degree programmes during the 1990s compared to that of the past decades. But my own experience in interviewing candidates for doctoral programmes, faculty positions, or other jobs over the last several decades as also the feedback from a large number of colleagues and friends, makes me believe that *on an average* the quality of degree holders - be it bachelor's or master's or a doctorate degrees - has perceptibly declined over the years. There are, of course, exceptions. We still have some institutions who have maintained their peak level of quality and we still have outstanding graduates coming up now-and-then. But the general output seems to have deteriorated. This deterioration is in the following terms.

Lack of clarity and depth Ever so many incumbants answer to questions with, 'I know the answer sir but at this moment can't express it ...' or 'it is on the tip of my tongue ...' or 'I have studied this topic sir and if given an opportunity I can handle it ...' They become apologetic as soon as a question is put to test the clarity of their understanding. In general the grasp of the fundamentals is poor, comprehension is blurred, and there is a general vagueness and tentativeness in whatever they say about the fields in which they are supposed to have specialised.

POOR COMMUNICATION SKILLS

One can condone the discomfort of a youngster who is facing an interview board by attributing it to nervousness. One can also attribute poor oral communication *vis a vis* scientific or technical matters to lack of command over the subject, mentioned in the preceeding para. But even when one is dealing with simple day-to-day matters one comes across a shockingly large number of degree holders who say what they do not mean and take meanings which are different from what they were told. The situation with respect to *written* communication is still worse. Rare is the research sholar who can write even a *correctable* draft of his/her thesis. Generally what the students write of their research is so atrociously bad that it is beyond correction. As a result the guides end up writing all the theses of their students.

LACK OF ORIGINALITY

When the basic attributes of clarity and communication skills are missing in most, the more prized talent - originality - is naturally a rarity.

What are the reasons of the deteriorating quality of our average graduate? It can not be lack of opportunities or infrastructure because over the years teaching aids such as slide projectors, overhead projectors and video films have been increasingly used to *assist* understanding. The ubiquitous xerox machine, and the ever falling cost of photocopies, has made it a lot easier for a student to gather material for study or research than it was earlier. The concept of inter-libeary links which was in its infancy in the 1970s has become widely operational now. Students are, therefore, no longer forced to limit themselves to what is available in the libraries of their own colleges/

universities. *Current contents* enables one to know about all the papers published in one's field as soon as the primary journals come out. Earlier we had to wait for 3-6 months till the journals arrived by surface mail to know what was published. And as no single library could afford to purchase all the journals, one had to travel here and there to complete the primary literature search. Now we also have computerised literature searches available which enable a student to do in a day what his/her seniors needed months to accompalish. PCs are everywhere and so are all kinds of software packages illustrating concepts and assisting in calculations. India being still quite away from the international copyright net, these packages - pirated -are available to Indian students practically free of cost. With friendlier and friendlier word-processing packages coming every six months, writing a paper or a report has become infinitely less labourious than it was till a decade back.

On top of it we have UGC - sponsored educational programmes and a whole taxa of other audio-visual programmes developed by agencies devoted to distance education such as IGNOU, NCERT etc, have become available to assist our learners.

Then why the general standard is going *down* instead of going *up*? I am hard put to offer an explanation except that the phenomena is a reflection of the general erosion in the value system. 'Smartness' - which in popular perception is the ability to achieve, carrier-wise, the most with the least effort - has become a greater virtue than excellence. Majority of the students do not believe that their job prospects are directly proportional to their abilities. The student who takes the long and dull-looking path of grinding through books and journals is regarded by his peers as archaic — a *moron*. The one who can 'talk his way through' any situation - with least preparation, of course - has become the ideal for most.

As today's student is tomorrow's teacher we can easily see why the rot is going on unabated. The plain fact, on the other hand, is that the job propospects *are* proportional to ability. Selection committees *do* look for virtue, and are frustrated when they find it missing.

b) Problems Specific to Environmental Education

We have mentioned earlier in this write-up that the allecompassing nature of environmental studies is a mixed blessing. On one hand environmental studies provide the near-ultimate junction of disciplines - a zone where all other disciplines, coming from their separate ways, meet and interact. On the other hand the sheer crowding that results can lead to great confusion too. We had cited the example of Salim Ali School of Ecology which began as a confluance of numerous disparate disciplines but become unviable to such an extent that it had to be trifurcated into three less multi-disciplinary units.

If we view this special attribute of environmental studies in the context of the *general* problems our educational system is facing, we can visualise what is, and to what degree is, wrong with the environmental education in India.

CURRICULUM

What topics must be included in a programme leading to MSc in environmental science/ environmental studies? Ideally everything that adds upto environmental studies. But as this everything is *practically everything* most of the curricula end up having a little of everything, As a result the end product is often not sure of *any*thing. Industries need environmental scientists who can sample and analyse air/water/soils for them, or can do EIA (environmental impact assessment) for them. But most MSc curricula - bursting at the seams as they are in trying to cover all relevant topics - hardly have space for anything more than a paper each on environmental analysis (EA) and EIA. And that much exposure, given the fuzzy nature of our teaching-learning process — is hardly adequate to prepare an average student *actually* do EA or EIA. Our MS (Ecology) students are taught mathematics, statistics, and computer programming *along with* numerous other topics. But they emerge none the wiser in their grasp of maths, stat, or programming. When our average MSc (statistics) is ill at ease if asked to apply statistics to a real-life problem, what command an average MSc in environmental science or ecology will have over statistics by virtue of his/her limited exposure to the subject through a single paper?

Public and private sector organisations <u>also</u> need professionals who can design and operate their waste treatment plants or can take charge of industrial safety and risk management.

But these are niches meant for environmental/chemical engineers. Wildlife parks and zoos need personnel trained in wildlife management. There *are* programmes which train personnel in this field but environmental science/ecology degree courses are not among them.

ENVIRONMENTAL EDUCATION AND COMPUTER EDUCATION

We shall come back to a comparison of environmental science and computer science - two young disciplines of identical age and importance. The comparison can give us some more insights into what is wrong with environmental education and, hopefully, provide hints on what *can* be done to remedy the situation.

In its first few years computer science was developed by persons formally trained in electronics/electrical engineering, operations research, mathematics, statistics, and physics. All these happen to be closely related physical science disciplines. Even before computer science emerged as an independant discipline, the above mentioned contributory streams were liberally drawing from one another and helping each other to grow. Some other disciplines have also contributed, especially to computer hardware, such as materials science but these contributions have come in widely-spaced quantas and are rarely specific to computer science (developments in material science influence <u>all</u> types of instruments, not computer hardware alone). On the other hand environmental science is a heterogeneous mix of a large number of disciplines which are either unconnected or have, at best, only tenuous links.

The student intake in computer science/engineering programmes is also much less heterogeneous. MSc, BTech, and MTech programmes take students coming from a few closely related disciplines. MCA and PGDCA programmes are relatively more liberal, yet are far from the all-pervasive liberalism of environmental science intake. Then there is a focus inherent in computer science programmes with clear objectives of developing competence in software and hardware technology. The employment niches which the degree holders occupy are also more clearly defined. Apart from degree programmes and recognised diploma programmes, there is a surfiet of private institutions offering all manners of short-term certificate and diploma courses. Persons who take these end up as mere computer operators capable, barring exceptions, only of running well-known software packages or word-processing. But these 'professionals' too occupy clearly defined slots of what we can call 'computer clerks' in shops and small business houses.

The better focus of its educational programmes and the clear links between the levels of training and the employment slots has made computer education into a much 'harder' science than environmental education.

CONFUSION BETWEEN DEVELOPMENTAL PRIORITIES AND COMPULSIONS OF CONSERVATION

This paper may not be complete if we do not make a mention, albeit a passing one, of the confusion we so often see in the minds of most of our environmental science degree holders on the issues of optimal development and optimal conservation. Which developmental activity is harmful, and to what degree? Should the construction of a highway be opposed because it is going to cause cutting of, say, 1500 trees? If so what about the enormous emount of energy we waste, and the pollution we generate, by resorting to old, narrow, and circuitous roads? Are the trees that were going to be cut irreplaceable? Are, *all* things considered, the environmental costs of the highway *really* more than the environmental benefits? Are the social benefits of no consideration?

Every environmental science/ecology student has been told that big dams or thermal power plants cause adverse environmental impacts. If we ask, 'in that case what should we do?' Pat would come a reply, 'we must go for alternative energy sources such as biomass, micro-hydal, biogas etc? If we probe further we would learn that our up-and-coming expert has never paused to think what would happen to environment if we try to generate 1000 megawatt of electricity through biomass. He or she would never have thought that it will require thousands of square kilometers of prime land and huge volumes of precious water to grow the trees needed to generate the same amount of power that is generated within a few square kilometers area by a NLC or a TTPS. Nor would he or she have budgeted that converting this biomass to electricity will generate the <u>same</u> quantity of pollution, if not more, as would a coal-fired or lignite-fired power plant. And of course it is unlikely to have been reckoned that forest tracts and slow-growing trees that come in the way of a fast-growing biomass plantation would have to be cleared-off in order to generate the wood needed for our biomass-based power plant.

One of the authors (SAA) has often asked my ecology students, 'tell me, in your own homes would you like to put an LPG cylinder in your kitchen or would prefer maintaining three cows and a bedroom-sized biogas plant for your cooking energy needs?' They reply - not without embarrassment - 'an LPG cylinder'.

When someone passionately speaks against the so called ills of eucalyptus and the criminal acts of land-owners in converting garden lands into eucalyptus plantations, I ask - 'tell me if you are the owner of a few acres of land with responsibilities to earn the daily bread for your family, educating your son, and marrying off your daughters; if you have to pay off a debt of a sizeable sum; would you still grow a polyculture of ecologically benign but world-wise worthless trees on your land or plant the *most* remunerative species?'

We have cited these examples to further emphasize the lack of balance and focus in most of our environmental education programmes, as also the inevitably superficial and confused product they achieve.

WHAT NEEDS TO BE DONE

In our opinion we have to begin restructuring our environmental science/ecology/environmental studies programmes so as to produce, instead of generalists, professionals whose forte is environmental botany/environmental zoology/environmental chemistry/environmental economics/environmental sociology etc. Within these, relatively specific, curricula should be provisions for training students in specialisations such as forest ecology, limnology, marine ecology, environmental analysis, pollution studies, environmental toxicology etc. There is also a scope for producing professionals for manning specific and well-defined slots as is the case with the different levels of computer science education.

The relatively better position in which environmental *engineers* are placed bears out this point, Environmental engineering programmes have clear and specific orientation towards air/water/solid waste *treatment* and environmental impact assessment. The degree holders are trained to man well-charted territories. There are, of course, the usual problems of dwindling calibre that we believe is omnipresent in our education system today but environmental *engineering* education does not suffer from the kind of confusions that are endemic with environmental *science* education.

We must conclude by stressing that our efforts to uplift environmental education would have a greater meaning if they are integrated with a general thrust to improve the quality of *all* our educational programmes.

10

ECOLOGY EDUCATION - SOME NEW PERSPECTIVES *BASED ON FORECASTING*

A normative-intuitive forecasting study, based essentially on Seth-Harva technique of scenario building, is presented on the future of ecology education. 'Crisis Point' in the form of a feedback loop and 'point of fulfillment', also consisting a feedback loop, were identified from several alternative scenarios and an action plan was developed. Panel approach and brainstorming methodologies were also used.

In 1987 Pondicherry University started what till today is the only academic programme in India leading to a masters degree in ecology. Several universities have been offering, since approximately one and half decade, post graduate diploma and degree courses in environmental science which are distinct from the MS (Ecology) course either in their emphasis on environmental pollution or in their leaning on one or other sub-branch of environmental science. The MS (Ecology) course is more traditional in its name - ecology being recognised formally as a branch of science nearly a century before the birth of the term 'environmental science'. The course, however, is rather modern in its conception and the manner of implementation.

It is open to students who have graduated in any branch of study; interest and aptitude in ecology are the main prerequisites, which are ascertained through a competitive examination and interview.

The first semester provides intensive exposure to physical, biological, and analytical sciences contributing to ecology and in subsequent semesters majorecosystems are taught with strong emphasis on solving real-life problems through field work and quantitative approach.

The study presented here was undertaken by the author with assistance from the first and second batch of MS (Ecology) students. The study used the methodologies of panel approach, brainstorming, and Seth-Harva system of scenario construction.

Panel discussion helped in obtainig consensus on the basic premises of this study. The points of consensus were :

a) **Goal or Norm of Ecology Education**: It was agreed that irrespective of individual or organisational aspirations, and shorn of flicks, appendages, and externalities, the end-objective or the ultimate goal of ecology education is achieving ecosystem stability, in other words restoring the presently degraded ecosystem and conserving the healthy ones.

b) **Identity of Goals of 'Ecology' and 'Environmental Science' Education**: It was agreed that irrespective of differences in the basic structure, breadth, depth or stated objectives of various courses, the end objectives of the ecology course defined above is completely identical to the environmental science courses. Therefore the conclusions of the present study, drawn with respect to the MS (Ecology) course, should apply equally well to the environmental science courses.

The brainstorming sessions which were conducted intermittently with the scenario building exercises produced ideas which, when subjected to cross examination and careful feasibility analysis, did not seem to lead to vistas worthy of further exploration. Some of the ideas emerging from brainstorms are given in appendix I.

The key findings that emerged from the application of

the Seth-Harva scenario building methodology are summarised below. The survey resulting from the panel approach, described above, constituted the Step I of the Seth-Harva system. The subsequent steps are as follows.

IDENTIFICATION OF 'CRISIS POINT'

With the norms set, the 'crisis point' or, to express it more accurately, the 'crisis engine' was quickly identified as the feedback loop presented in Figure 1. Resource constraints, which result from population and development pressure (planned or unplanned) lead to environmental degradation which, in turn leads to resource constraints. Missing from this sub-model is 'technology constraint' or 'know-how constraint' because we felt that lack of technology or know-how is not a real contributor at all. The know-how to deal with sanitary, agricultural and industrial wastes is available, so is the know-how to prevent degradation or achieve conservation but it is the lack of resources, in precise terms monetary resources, which prevents the utilisation of the know-how. In such a situation the cycle, or 'engine' depicted in figure 1 keeps running.

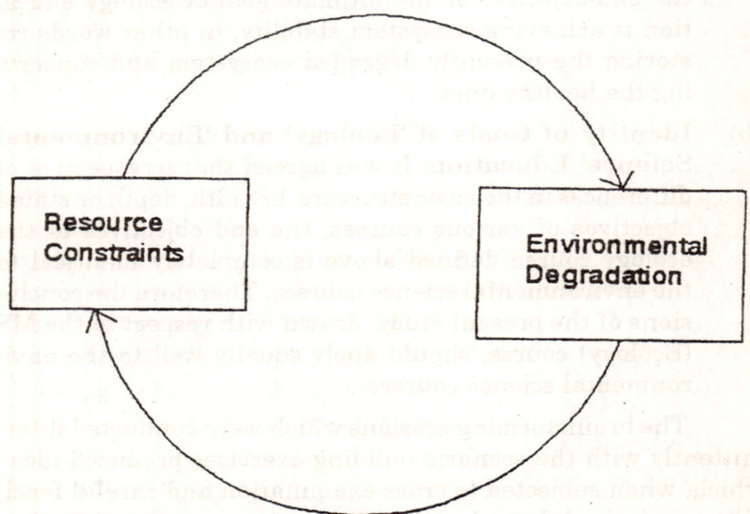

Figure 1 : The presently operating feedback loop: the 'crisis engine'

IDENTIFICATION OF THE 'POINT OF FULFILLMENT'

The norm 'a' set through the panel approach was identified as the ultimate point of fulfillment or the goal; it was considered that numerous spin-offs will emanate from this ultimate point of fulfillment, such as better social party and better quality of life.

DEVELOPMENT OF PREFERRED FUTURE SEENARIO

Several alternative scenarios were constructed and scrutinised. The one we preferred to form the basis of the subsequent step of action-plan development has, as its key component, another feedback loop, depicted in Figure 2.

The preferred scenario for achieving the desired goal consists of utilising human resources as the input to environmental conservation. Several justifications for preferring this scenario were apparent :

a) human resource is our very strong point; we have abundance of it and not shortage is likely in the next 50 years at least.

b) every human being interacts with the environment irrespective of his/her social, economic, or educational status and, baring a very few, all are directly effected by environmental quality. Upgradation and conservation of environment, therefore, should be to everyone's benefit; creating awareness of it should facilitate human resource mobilisation.

c) The task of restoration/conservation is so enormous, and the overall situation is such, that the possibility of generating monetary resources to accompolish this task are nil atleast in the next 50 years.

d) The population pressure which has triggered the cycle depicted in figure 1 would continue to generate new stress on the the environment (and resources) unless the cycle is reoriented to reverse the trend. In other words even if we provide one-time massive caapital injection for restoring the environment the 'crisis engine' will quickly annule the restoration.

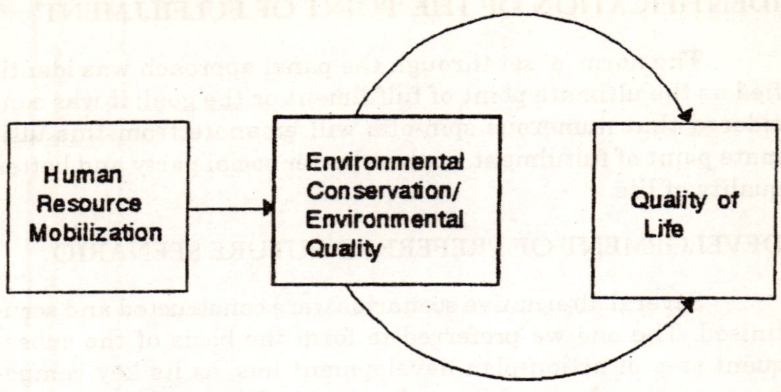

Figure 2 : Main components of the preferred scenario

THE ACTION PLAN

The action plan was worked out vis-a-vis ecology education. While the *end objective* of the ecology education appears achievable only through human resource mobilisation, the present syllabi *incorporate no consideration to this aspect at all.* Therefore mass participation is the solution but is it a utopia? Is it not a fancy idea that will be swept aside by the bitter realities of human strife?

The answer to this apprehension came from the models of Lake Biwa and Chipko. In the former case a dying lake in Japan and in the latter case a doomed forest in India were revived solely by mass participation at the grass-roots level. Therefore it is possible; only we have to find ways and means of doing it on a larger and wider scale.

The action plan consists of integrating an innovative formal course and batch-to-batch linked project *with the existing two-year masters programmes.* In essence the plan consists of -

a) Introducting a novel three (or four) credit course, 'Human Resource Mobilisation for Environmental Management'. The course will orient the students towards the basics of mass communication, with stress on methods which would generate confidence of the target population in the usefulness as well as *sincerity* of environmental conservation measures. The course *must* not project elitism, nor generate a feeling that some elites are teaching the common folk on the 'dos' and the 'donts'. Rather it must create a feeling that environmental conservation is essential for everyone; that *everyone* stands to gain by environmental resotration and protection. It should rely on approaches that people at the grass root level should find easy to comprehend; the approaches should, so to say, touch their heart.

b) Integrating the formal course with a two month summer project. In the summer project the students will go to a locality and actually join the people in restoring that locality. Here again the approach will not be of teacher and taught or of an elite grooming the uninitiated. It will be rather of working in tendum, working in step.

c) It was felt that the formal course should come in the second semester - which runs from January to April - followed by the summer project. Once the summer project is over the students going into the third and fourth semesters will maintain continuity of the field-work by taking turns to visit the locality and assist the people.

d) It was also felt that to have a worthwhile impact it may be necessary to identify one or two localities and work extensively in them till substancial results are achieved.

e) It was considered that there may be voluntary agencies or governmental agencies also operating in the chosen locality. The question of the student's role vis-a-vis these agencies was critically examined. It was felt that the students will actually try their best to invoke the feelings of identity of purpose and cooperation in them rather than stepping on their toes or alienating them. In fact winning the affection and confidence of these agencies will be the very first task of the students.

f) We were well aware that the task will be made complicated by vested interests and antagonists. The students must learn to involve people and mobilise cooperative efforts *inspite* of these forces. In fact, it was recognised, that this will comprise the real training as this is the single most important constraint in environmental conservaiton.

g) The continuity will also be maintained by the faculty co-ordinating the student's project.

As the proposed modification does not drasticaally alter the present course structure, nor does it increase the time required in completing the degree,it is likely to be acceptable and it should be possible to implement it with minimal teething troubles throughout the country.

LIKELY IMPACT OF THIS ACTION PLAN

The following impacts were recognised :

i) The proposed programme will expose the students to the *real* real-life problems. Some of the students may later take up activist role but even those who join the conventional career streams will have had a first-hand knowledge of the problems at the mass level and would have gained insights on how such problems can be addressed. The exposusre will therefore have a strong influence on the policies, programmes and actions to which these students will contirbute as technocrats scientists or educators.

ii) The ecology students having been groomed (at least partly) in the theory and practice of ecology should be able to perceive the grass-root problems and more innovative and professional way. For the same reasons their actions may carry more convictionthan the actions of well-meaning, honest, but lay conservationists working in the same locality.

iii) The student's projects will not merely be an exercise in improving a small area or locality. They, rather, will help in censitising and inspiring 'collectivism towards conservation'. They will also generate much needed expertise for others to use.

iv) In quantitative terms, the impact of this action-plan over a 50 year time span was visualised to take the sigmoid form depicted in Figure 3. A gestation period of five years, after which the impact will become pronounced as a result of multiplier effect. A shaarp rise and then the logical culmination of near-total restoration in the form of a plateu. It may be described by the equation:

$$Y = + \frac{K}{1 + e_a - rt}$$

Where Y is impact magnitude on a 0-10 scale; t is time in years; and k, a and r are constants.

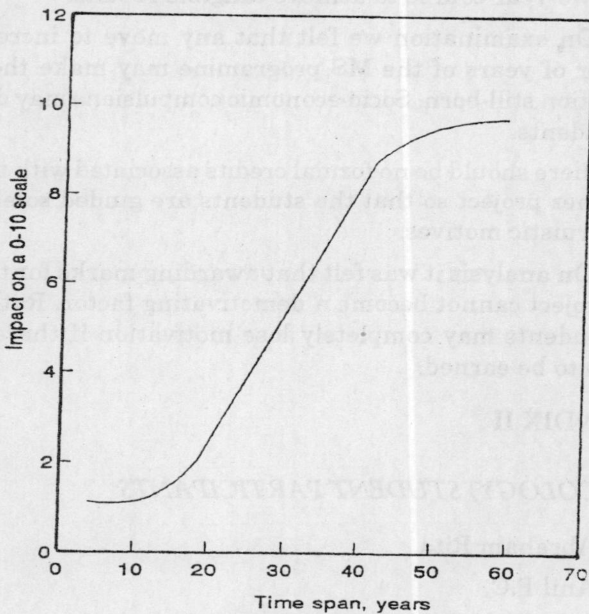

Figure 3 : Anticipated impact of the action plan

ACKNOWLEDGEMENT

Author is grateful to MS (Ecology) students for providing the initial entusiasm. The author was fortunate in getting this manuscript read by eminent ecologists during the last one and half months - Prof Belsare, Prof Gopal, Prof Fred Perry, Prof Harte. Their encouragement was extremely gratifying. Many thanks to Tanseem and Tarannum for patience and understanding.

APPENDIX I

Some of the ideas put forward during the brainstorming sessions are summarized below :

a) A full year needs to be devoted in the field after the present two-year course to achieve tangible results.

On examination we felt that any move to increase the number of years of the MS programme may make the entire innovation still-born. Socio-economic compulsions may dissuade the students.

b) there should be no formal credits associated with the summer project so that the students are guided solely by altruistic motives.

On analysis it was felt that awarding marks for the summer project cannot become a demotivating factor. Rather several students may completely lose motivation if there are no credits to be earned.

APPENDIX II

MS (ECOLOGY) STUDENT PARTICIPANTS

1 Abraham Rita
2 Anil P.C.
3 Anitha S.
4 Aparna T.G.
5 Ganesh K.B.
6 Ganesh T.
7 Kinhal Vijayalakshmy
8 Lal Rashmi
9 Mathew A.G.
10 Premdas Sharmila
11 Rai Devadas Nitin
12 Ramesh N
13 Rao Namita
14 Rao Ragavendra K.

15 Ramakrishna L.
16 Roy Pratim
17 Ravi N.
18 Rao Sunita
19 Santharam V.
20 Santosh S.A.
21 Sinha Aditi
22 Sengupta Nina
23 Somanathan Hema
24 Shylaja M.N.
25 Soubatra Devi
26 Udayan Niraj Joshi
27 Uma R.
28 Uma G.

11

FORECASTING OPEC OIL PRICE HIKES - *DIAGNOSING FAILURES AND EXAMINING CRITERIA FOR THE FUTURE**

Techno-economic forecasters have achieved several notable successes over the past decades, but they have also suffered the mortification of some forecasts going wide off the mark. The most poignant 'miss' was the total failure to forecast the oil prices hikes of 1973 and 1979. Equaly sobering was the failure to forecast the oil price crash that occurred in the 1980s. Both these events - the bust and the boom - had connotations of windfall / catastrophe for OPEC/non-OPEC countries depending on who was at the receiving end. The events rapidly altered the global energy scenarios and caused sudden spurts and sudden breaks in many areas of research and development. We have attempted an examination of the possible factors that led to the mis-forecast. We have also worked out the factors which will play a dominant role in the 'energy game'; the factors which must form the basis of the scenarios for energy futures.

INTRODUCTION

Between 1960 to 1973 the world oil prices were hovering

* With Naseema Abbasi, P.C.Nipaney, and E.V.Ramasamy

between $ 5.4 to $ 7.0 per barrel (Figure 1). In 1973 OPEC countries flexed their muscle. There was no technological breakthrough or failure. there was no real shortfall in production and there was no sudden jump in demand. The trends for these remained what they were between 1960 to 1973. Yet the oil prices were unilaterally raised by OPEC countries so that by the end of 1974, the cost of oil had risen to three times its 1973 value. Another, more sharp, hike was made in 1979 so that in 1980 oil was selling at eithteen times its price of 1973.

The forecasters, who were taken aback by these hikes were now projecting that the oil price will continue to rise and the 1990 price will be double the 1980, price.

Instead there came a crash. Instead of the anticipated rise $ 40 per barrel by 1986, the prices plummetted to $ 10 per barrel. What had happened?

WHAT WENT WRONG

The 1973 hike was so devastating that it created a warlike situation for non-OPEC countries. Economic compulsions forced the non-OPEC countries to take to energy conservation. It also gave a spur to exploration and indigenous production of oil. As the prices were high, indigenous production brought huge benefits which led to more exploration and production. The net result of energy conservation was decrease in the demand for oil. The net result of increased non-OPEC oil production and more reliance on alternative sources of energy was an increase in supply. The impact of these reached a climax soon after the second OPEC price hike of 1979. The impact pulled the prices down; they began to fall. To check this fall OPEC decided to reduce production and assigned fixed production quota to the members (Figure 2). But the need for petrodollars made it difficult to reduce production. Consequently the production quota were exceeded by many OPEC members. The ceilings on production were becoming so meaningless that they had to be lifted. Ultimately the prices crashed to $ 10 per barrel by mid-1986 (Figure 2).

This rise and fall was neither perpetuated by the growth or failure of new technologies, nor by any catastrophe (such as the 1991 Gulf war) which had suddenly increased demand or reduced supply. The rise was triggered solely by the OPEC power and the fall came when this power was successfully eroded.

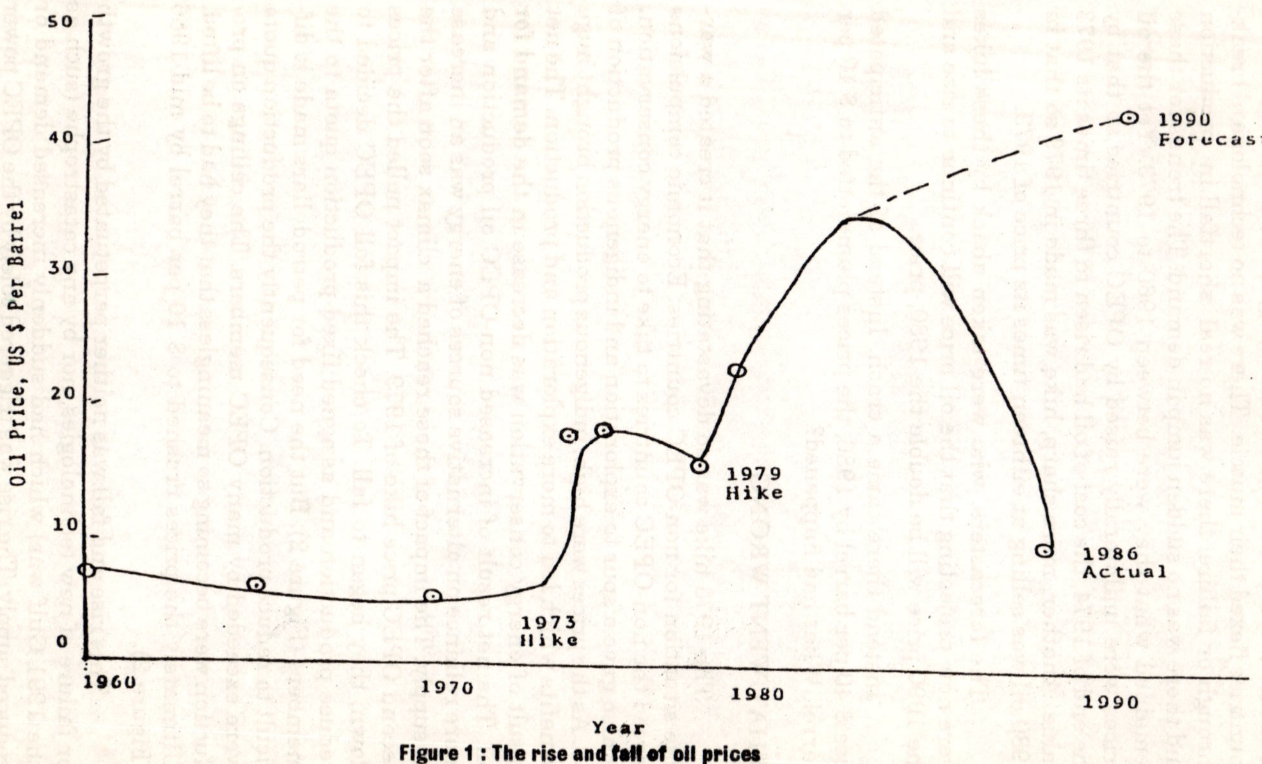

Figure 1 : The rise and fall of oil prices

THE MIT FORECAST : HOW PRECISE?

Let us now look at one of the most well-considered set of forecasts made in the past decade. The year of forecasts was 1986 which was three years <u>after</u> the first oil price hike, three years <u>before</u> the second oil price hike, and a decade before the oil price crash MIT, 1977. The forecasts, based on the analysis of several carefully constructed scenarios had the following key components:

1. The supply of oil will fail to meet increasing demand before the year 2000, most probably between 1985 and 1991, even if energy prices rise 50% above current levels in real terms. Additional constraints on, oil production will hasten this shortage, thereby reducing the time available for action on alternatives.
2. Demand for energy will continue to grow even if governments adopt vigorous policies to conserve energy. This growth must increasingly be satisfied by energy resources other than oil, which will be progressively reserved for use that only oil can satisfy.
3. The continued growth of energy demand requires that energy resources be developed with the utmost vigour. The change from a world economy dominated by oil must start now. The alternatives require 5 to 15 years to develop, and the need for replacement fuels will increase rapidly as the last decade of the century is appproached.
4. Electricity from nuclear power is capable of making an important contribution to the global energy supply although worldwide acceptance of it on a sufficiently large scale has yet to be established. Fusion power will not be significant before the year 2000.
5. Coal has the potential to contribute substantially to future energy suplies. Coal reserves are abundant, but taking advantage of them requires an active program of development by both producers and consumers.
6. Natural gas reserves are large enough to meet projected demand, provided the incentives are sufficient to encourage the development of extensive and costly intercontinental gas transportation systems

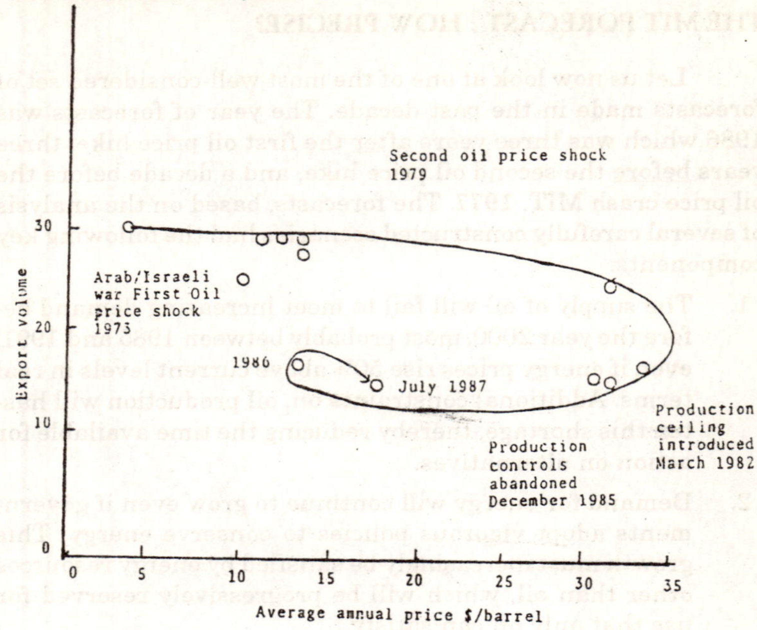

7. Although the resource base of other fossil fuels such as oil sands, heavy oil, and soil shale is very large, they are likely to supply only small amounts of energy before the year 2000.

8. Other than hydroelectric power, renewable resources of energy such as solar, wind-power, wave-power-are unlikely to contribute significant quantities of additional energy during this century at the global level, although they could be of importance in particular areas. They are likely to become increasingly important in the 21st century.

9. Energy efficiency improvements, beyond the substantial energy conservation assumptions already built into our analysis, can further reduce energy demand and supply. Policies for achieving energy conservation should continue to be a key element of all future energy strategies.

10. The critical interdependence of nations in the energy field requires an unprecedented degree of international collaboration in the future. In addition it requires the will to mobilize finance, labour, research and ingenuity with a

common purpose never before attained in times of peace; and it requires it now.

It is evident that except the first forecast, others are very close to the bull's eye. In fact misjudgment in the matter of oil price has hardly clouded the other forecasts. The reason behind this, yet agin, is that the rise and fall of oil prices was not a consequence of technology but external political/international forces. In terms of assessing the growth of various technologies and alternatives, the forecasts are almost uncanny. Most noteworthy is the accuracy of the forecasts that nuclear power and renewable energy sources such as solar will not contribute substancially to world energy production in this century. These forecasts were made at a time when renewable energy sources were being considered a penacia by the laymen and when long Island or Chernobyl had not happened. In fact in 1976 nuclear power was considered to have 'most outstanding safety record'. Yet the professional forecasters were successful in perceiving their limited role till atleast 2000 AD.

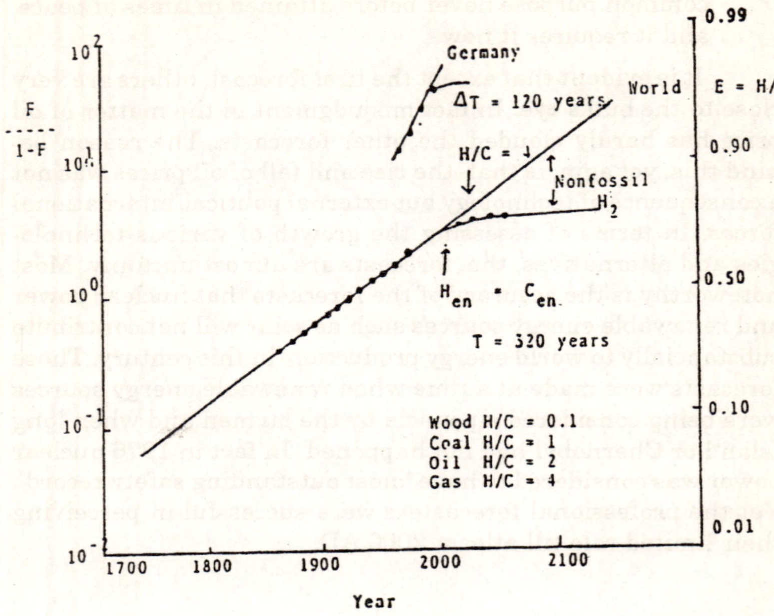

Figure 3 : Hydrogen-to-carbon ratio in the World's fuel mix

LESSONS FOR FUTURE

The main lesson from the failure in forecasting oil prices is: do not rest content with the accurate assessment or various technologies but keep an eye on the nations or group of nations who are either major producers of energy or who possess one or other major energy resource. They are the main actors in the game of energy pricing. These actors have the power to change the complexion of the game. While technological breakthroughs will have their sure impact the actors can bring about devastating changes in the global energy scenario even when the technologies remain what they are. In short, it will be dangerous to put all eggs in the OPEC basket.

Other factors which will play a dominant role in shaping future scenario are :

a) *Economic recovery* : it will revive demand for oil - especially in developing countries. It is happening in case of India.

b) *Limits to conservation*: improvements in energy efficiency and conservation would reach their limits of saturation. Large-scale impacts in cushioning energy demand from these factors are unlikely to come by in future - though major break-through in apparently unrelated field of superconductivity may again push up the limits achievable in energy efficiency.

c) *Limits to non-OPEC oil Production* : The oil production of non-OPEC producers has reached a level which may be difficult to sustain in the long-run. The fallen oil prices will also discourage oil exploration and processing the long run.

d) *The undeniable 'monopoly' of OPEC*: It is a fact that, oil hike or oil collapse, two third of world's proven oil resources are located within OPEC members. Thus an OPEC recovery can take place any time if not with the same crushing impact as was in the 1970s.

e) *The continually rising hydrogen to carbon ratio*: Between 1888 to 1988 the hydrogen to carbon ratio of the fuels has continually risen. The trend has been impeccably linear for nearly a century but is tending to break upwards now (Figure 3). This has portants of hydrogen energy being the major fuel of the twentyfirst century.

REFERENCES

1. Massachusetts Institute of Technology, 'Energy Global Prospects 1985-2000', Report of the Workshop on Alternative Energy Strategies held in 1986, MIT, Massachusetts, 1977, 45 pages.

12

THE STATE OF INDIA'S *ENVIRONMENT - A QUICK LOOK*

Three years from now we would be entering into a new century as well as a new millennium. In three years we will also be reaching a stage where our population would cross 1 billion and the world's population 7 billion.

In the coming three years we would continue to witness spectacular growth in science and technology. It is hoped that by 2000 AD most of the educated Indians would have a cellular phone using which it would be possible to talk to anyone in the world from anywhere in the world. Computers would so much shrink in size and so much advance in capability that we will be able to carry in our pocket a computer which will be capable enough to do most of our secretarial work.

In the coming three years we will also witness a rapid decline in the availability of land and water, food and fibre, hearth and shelter. The air, water, and land will also rapidly deteriorate in terms of quality, unable to assimilate the pollution caused by the burdgeoning billions.

The recent past unveiled the spectre of ozone hole and greenhouse effect. In the coming future we may reap newer harvests resulting from the excesses that are being committed on the environment.

Can we save the earth? Can we save ourselves? Is it necessary to save the environment if we want to survive?

The environment is not just pretty trees and tigers, threatened plants and ecosystems. It is the very entity on which we all subsist, and on which our entire agricultural and industrial development depends.

India is unique in the context of its environment. This is because we have three key elements of nature in abundance. The first is sunlight, and we have lots of it. The second is water and again we have lots of it (yes, India is one of the best endowed countries in terms of rainfall), if only we can learn to hold it properly. And then we have excellent soils. These three elements, when combined properly, become the fountainhead of life. In fact, according to FAO, if there is any country in the world which has the potential to feed half of Asia's population today, it is India!

But if not combined properly, these same elements can become harbingers of death. The high sunlight and high temperatures can desiccate the soil, and the high rainfall can wash the soil away. And in a country like ours, experts say this degradation can take place at a rate faster than any other part of the world.

According to available official figures, between 100 to 150 million hectares of land area is rapidly turning into desert. Every year 2.5 mha - that is, about one percent of India's land area - turns into wasteland as a result of deforestation, mining, waterlogging, and soil salinization.

Then, on an average, every hectare of land loses 20 tonnes of topsoil every year. Indeed as much topsoil gets washed away every six months as is used to build all the brick houses across the country!

WATER POLLUTION

It is said that 70 per cent of all the available surface water in India is polluted, though one may wonder where is the remaining 30 per cent! There seems to be no clean water outside the mineral water bottles! And even that is not *always* clean! Shokingly, even the high altitude lakes are becoming grossly polluted.

The condition of groundwater is no more reassuring. On the hand we are cutting forests and reducing recharge of the groundwater reservoirs. On the other hand we are pumping out large quantities of groundwater.

The water pollution problem is getting alarming all over the country. If a river is not chocked by industrial effluents, it is reeling under the onslaught of sewage and other domestic wastes. More often than not water bodies are getting the worst of *both* types of pollution, plus harmful run-offs laden with pesticides and fertilizers from agricultural fields! Pictures of effluent waters from chemical factories are often so full of colour and muck than no laboratory test is needed to proclaim their toxicity.

AIR POLLUTION

Air pollution levels in most India cities are going up, up, and up. In several cases they have begun exceeding recommended limits long ago. The Bhopal disaster, followed by recurring instances of poisonous gas leakes and industrial explosions has made us further aware of the high risk environment we live in. Normally when we speak of air pollution we think in terms of big cities like Bombay, Calcutta, Delhi and Madras but the contributions from vehicles and chulhas all over India match the pollution caused by industries.

We often think that the problem will be solved by pushing industries into backward areas. But once an industry is set up in a backward area, there are almost no controls on it. For instance, the Patratu thermal power station, 70 kms away from the town of Ranchi, belches smoke. But no one has even talked of controlling pollution from this power station.

And what of *indoor* air pollution in the majority of Indian households? It has been estimated that the pollutant levels *and* the duration of daily exposure to them are often much higher than in the badly polluted industrial localities. Poor-quality fuel combined with burners of poor efficiency, and poor ventilation are responsible for this shocking state of affairs.

AN ACTION PLAN FOR ENVIRONMENT

Protecting the environment is now enshrined in our constitution as a fundamental duty of every citizen. But saving and enriching our environment is easier said than done; indeed it is such an enormous task that no regulatory agency, no governmental machinery, and no amount of monetary or technological inputs can achieve it *unless* each and every citizen contributes his or her mite. A few suggestions culled from the works of different environmentalists are presented below.

(1) Protection of natural resources from irrational pressure of development

We must protect whatever is left of our air, water, land and forest resources. This means that our various acts like the Forest Conservation Act, the Air Pollution Control Act, the Water Pollution Control Act and the new Environment Protection Act must all be strictly implemented overriding all political and business pressures. Individual citizens and citizens' groups must also raise public opinion and use the acts available to them.

(2) Envrionmentally-sound planning built around concepts of equity and sustainability

It is vital that we improve the standards of living of the vast majority of our people. Therefore, as important as protection, is to use our natural resources at a higher level of productivity and in a manner that the increased production is sustainable. In other words, we must simultaneously aim at growth, equity, and sustainability.

Sustainability demands that when we plan for the economic development of an area we simultaneously plan for the environment so that the growth can be sustained without harm to the ecosystem.

(3) Employment guarantee built around programmes for ecological regeneration

Ecological regeneration is a very labour intensive exercise, whether it be soil or water conservation programmes, digging ponds and tanks, or afforestation. This is good for us because once a decentralized ecodevelopment plan is available for a particular ecological region, massive employment can be created through activities aimed at the restoration of the ecological infrastucture. If we can tie employment generation with ecological regeneration we can decisively move towards meeting people's needs on a sustainable basis.

(4) Rural development must become equitous and must lead to holistic enrichment of village ecosystems

It just paraphrases what Gandhiji once said, that our villages must become self reliant in their needs if the nation has to become self relient.

(5) Resource-rational urbanisation

Managing the world's largest urban population, that is in India, poses immense challenges and the answer lies in technology choices built around the principles of equity and resource-relationality. For instance, can we think of improved mud buildings with an excellent telephone system; transport systems dependent on a conjenctive use of bicycles and modern buses; composting toilets with a fleet of vacuum suction trucks, etc.

(6) Conservation-oriented education

We must make our people environmentaly conscious. Educated, urban individuals are today so alienated from their environment that thery have become, what is called, 'resource illiterate'. But as powerful and heavy consumers of natural resources they can do enormous environmental harm.

(7) A national rehabilitation policy

If dams must be built and if industries and cities must expand, we must develop a national rehabilitation policy for those who are displaced as a result of development. This proposed policy must make resettlement exercises honest and open; respectful of people's economic and cultural needs and their demand of land for land. Development must not be allowed to become a cause of impoverishment for anyone.

Mahatma Gandhi had said, 'there is enough in this world for everybody's need but not for everybody's greed'. Burdened though we are, we can still survive if we build our development strategies keeping these golden words in mind.

PART- II

TUTORIALS

13

WATER - VERY SPECIAL VERY PRECIOUS

This tutorial celebrates the uniqueness and pricelessness of water - as a 'wonder chemical' and as the single most important natural resource.

To say, that water is a precious resource may well be an understatement. Water is not *merely* a precious resource. It is *the most precious resources of all!*

For most of the life-forms, including us humans, oxygen is essential. However there are organisms which can live without free supply of oxygen (such as anaerobes), but there is not a single organism which can live without water. There is no life if there is no water. Water is that precious!

We have been supplied too, quite abundantly, with this resource (Table 1). We all know that two-third of the world is water, mostly in the form of oceans which contain 97% of all the water. The remaining 3% is freshwater. Here again the bulk of freshwater, three fourth of all the freshwaters to be precise, is in the form of polar ice and glaciers and about one fourth is beneath the ground, as groundwater. The rivers and lakes which we see and hear so much about contain less than one per cent of the total freshwater!

Table 1.

Distribution of water in the hydrosphere

	Volume (km3xlO3)	Percentaye of total vol.	Percent of fresh water	Replacement time (years)
Oceans	1370323	93.94180		3000
Deep groundwater	60000	4.11320		5000
Active groundwater	4000	0.27400	14.094	330
Ice	24000	1.64500	84.566	8000
Lakes	**280**	**0.01900**	0.987	7
Soil moisture	85	0.00580	0.299	1
Atmosphere	14	0.00096	0.049	0.027
Rivers	1.2	0.00008	0.004	0.031
Freshwater total	28380.2	1.92000		
Water, Ground total	1458703	99.99984	99.999	2800

However this small fraction of water is the most easily available fraction and has therefore been utilised to the greatest degree so far. It is quite another mátter that even this is not utilised in full due to several reasons which will be dealt upon subsequently. So far only less than half of the total water available in the rivers of India is utilised.

Where does the rest of the water go? In reality it does not go out of our grasp because it keeps coming back to us through what is called the 'hydrologic cycle'. The rivers drain into oceans from where the water evaporates into the atmosphere. The amount of water thus going up is enormous - some trillion tonnes per day. Each day this much water comes back on earth as rain, snow, sleet and hail. We use some of it and the rest goes back to where it came from. And this goes on and on thus the water which was once used by our ancestors is still there the water from which Archimedes jumped out naked shouting 'Eureka, Eureka', the water which boiled over in James Watt's Kettle and heralded the industrial revolution, and the water with which Mendel tended his peas to develop the laws of heredity - is still very much there. Only its molecules are now dispersed throughout the world. We may never know that some of the water that we would drink in our next glass may have played a key role in shaping the history of mankind!

And how much water do we need? It is a question really difficult to answer. While we try to find an answer we become all the more aware of the preciousness of this tasteless, odourless and colourless liquid. It is quite easy to quantify the direct uses of water and to say that an average Indian consumes about 200 litres of water per day compared to about 350 litres consumed by an average North American and 50 litres by an average African. However water is the heart and soul of many other things that we come across. From its creation to our breakfast table an egg requires 500 litres of water including the fraction of water required by the chicken and the various plants and animals eaten by the chicken which have provided sustenance for that one egg. A vegetable pulao has much as 14,000 litres of water behind it - water which was used to grow that much rice, to raise that much vegetables, to get that much spices etc. And the manufacture of the paper on which these words have been printed for you, has itself consumed anything from 15 to 30 thousand litres for its each tonne. Similarly enormous amounts of water is most essentially required to grow or manufacture or construct anything and everything which we use in everday life (Table 2). In short not only our own existence but the existenee of all other living and non-living things on which we depend, shall cease if there is no water.

RAINFALL AND INDIA

In terms of annual rainfall, India is one of the wettest countries on earth (Table 3), better off than such developed countries as USA, UK, Canada, Spain, and Sweden. Even in terms of *per capita* rainfll India, inspite of its large population is better endowed than such countries as UK, Germany, Italy and Japan. But, unfortunately, this rainfall is very unevenly distributed. Some regions get it too much too quickly, and some too little too slowly. Besides, as one would see from most of the chapters in this book, what we do receive, gets polluted.

WATER - THE WONDER CHEMICAL

The unique properties of water, which make life possible on earth, also play a decisive role in influencing the ecology of the water-bodies : lakes, ponds, rivers, oceans. An understanding of these factors is essential in getting to the 'depth' - literally and metaphorically - of the problems that arise in the water-bodies and in developing strategies of solving these problems for their effective environmental management.

Table 2

Water consumption by major industries

Items	Tonnes of water usage per tonne of production
Steel	100
High quality paper	310
Low quality paper	220
Oil	10
Petro-chemicals	1,010
Aluminum	190
Rayon	1,800
Acetate rayon	3,160
Cement	3
Glass	12
Chemical seasoning	1,680

THE UNIQUE PROPERTIES OF WATER

Water exhibits some unique properties, which derive from the structure of the water molecule and its tendency of form aggregates (Annexure I). Table 4 summarises some of the most important properties.

Eventhough water freezes at ~ 0°C it is not at its densest when it freezes - as most other liquids are - but exhibits its highest density at ~ 4°C (Table 5). Then water has exceptionally high latent heats of melting and evaporation and exceptionally low thermal conductivity (Table 4). All these properties have great significance in helping aquatic organisms in cold climates to survive when the lake or river or pond in which they live begins to freeze at the top. As ice is lighter than the water of near-zero temperature a layer of ice forms on top of such water bodies and stays there. Further the ice layer thus formed helps to insultate the underlying water due to water's poor thermal conductivity. As a result water bodies do not freeze all the way down even when air temperatures go well below 0°. This enables the aquatic organisms to survive.

When spring comes and the air temperatures go above 0°C, ice's high latent heat of melting prevents it to melt too fast. Consequently the process of melting is slowed down and the organisms dwelling in the water body are saved from the sudden shock of rising temperature.

When the temperature of water rises beyond 4°C its density becomes lesser with each degree of rising temperature. But this does not happen in a uniform fashion (Table 6). The net fall in density between 0°C and 5°C is lesser than that which occurs between 5°C and 10°C. This in turn is lesser than the fall in density between 20°C and 25°C. In other words a lake in the tropics in which the surface temparature is 32°C and bottom temperature 22°C there will be sharper difference in the density of top layer and that of the bottom layer than in a lake of a temparate region having surface temperature 22°C and bottom temperature 12°C.

Because water has poor thermal conductivity all deep water bodies - in which there is appreciable temperature difference between the top and the bottom - get 'stratified'. In other words strata or layer of water get formed, with warmer (hence lighter) water lying over cooler (hence denser) water. Such stratification is more pronounced in tropical water bodies because *difference* in the density of water increases as the temperature increases and the higher ambient temperatures of the tropics (compared to temperate regions) induce more pronounced density stratifications.

Water has exceptionally high latent heat of evaporation (i,e the heat required to evaporate water) - indeed higher than any other naturally occuring liquid; even higher than that of mercury. This ensures that even during tropical summer the lakes and reivers do not dry up easily. If the lakes were filled with mercury instead of water they would have dried up faster!

Other exceptional properties of water are its high dielectric constant and high surface tension. The former property enables water to easily dissolve a large number of organic and inorganic substances. This ability of water has brought for it the sobriquet of 'universal solvent'. Higher surface tension

Table 3.

Rainfall and per-capita water received in different Countries

Countries	Population x lo6	Area x 103 km2	Prescipiation per year mm/year	Total amount of precipitation per Syear 103m /year	Amount per capita per year m3/year,mm
Canada	2,283	9,976	522	52,075	228,099
USA	21,361	9,363	760	71,159	33,313
UK	5,600	241	1,064	2,564	4,579
France	5,291	551	750	4,133	7,31i
W.Germany	6,183	248	803	1,991	3,220
Italy	5,581	301	1,000	3,010	5,393
Spain	3,547	496	600	2,976	8,3°Q
Sweden	820	450	700	3,150	38,415
Austria	752	84	1,191	1,000	13,295
Swiss	640	41	1,470	603	9,421
USS2	25,430	22,345	502	112,172	44,'10
Rumania	2,118	238	700	1,666	7,366

Yugoslaiva	2,133	256	975	2,496	11,702
China	82,786	9,597	660	63,340	7,65
Philippine	4,183	299	2,360	7,056	16,868
Thailand	4,133	514	1,420	7,299	17,660
Indonesia	12,759	1,904	2,620	49,885	39,098
India	57,422	3,230	1,170	37,791	6,581
Iron	3,301	1,648	250	4,120	12,481
Saudi Arabia	891	2,150	Approx.100	2,150	24,130
Egypt	3,723	1,001	65	651	1,749
Kuwait	98	22	120	26	2,653
Australia	1,334	8,084	460	35,346	264,963
New Zeeland	**310**	**269**	2,010	5,407	174,419
Japan	**11,193**	**377**	1,788	6,749	6,030

enables water to rise very high in capillaries against the pull of gravity. Due to this property water is able to rise up to the peak of the tallest of trees.

Table 4.

Physical properties of water

Density (25°C) in kg/m3	997.075
Maximum density in kg/m3	1000.000
Temperature of maximum density °C	3.940
Viscosity in Pascal/secs (25°C)	0.890x10-3
Kinematic viscosity in m2/s (25°C)	0.89x10-6
Melting point °C (at 101325 pa)	0.0000
Boiling point °C (at 101325 pa)	100.00
Latent heat of melting in kJ/mol	6.0104
Latent heat of evaporation in kJ/mol	40.66
Specific heat capacity (15°C) in J/kg°C	4186
Thermal conductivity (25°C) in J/kg°C	0.00569
Surface tension (25°C), N/m	71.97x10-3
Dielectric constant (25°C)	78.54

Yet another property of water which plays a significant role in regulating life on earth is viscosity.

During movement of aquatic organisms in water, frictional forces come to play to an extent dependent on the velocity of relative motion, and shape of the body and also the viscosity or kinematic viscosity of the water. An organism which propels itself through warm water therefore uses up less energy than in cold, and at 25°C a primarily free swimming organisms sinks at twice the rate at 0°C, provided that the water is still. By the same token, the mechanical force exerted on stationary organisms in flowing waters can be either raised or lowered due to change in temperature. In addition the thickness of the laminar boundary on surfaces over which the water flows is also affected by the viscosity. The biological significance of the viscosity of water is still to be fully understood and the observations made here only indicate that there are interesting possibilities for further research.

The viscosity is the resistance offered by water during

Table 5.

Temperature, density, and specific volumes of water

Temp (°C)	Density (kg/l)
0 (ice)	0.91860
0 (water)	0.99987
4	1.00000
5	0.99999
10	0.99973
15	0.99913
18	0 99862
20	0 99823
25	0.99707
30	0.99568
35	0.99406

Table 6.

Temperature, buoyancy, and density changes of water

Temp (°C)	*Buoyancy (gig)*	*Density change/°C*
0 Ice	85.1 x 103	—
0 Water	132	7.4
4	0	—
5	8	1.0
10	273	10.0
20	1870	25.3
25	2929	31.5

* Density relative to that at 4°C

** Density change/°C relative to that between 4 and 5°C

free flow or other changes in physical form. It is the force required to diplace a mass of 1 kg by 1 m during 1 s in the medium. Its unit is the Pascal second (Pas) = 1 kg/m/s (1 Pa = 1 N /m ^{2}) the viscosity is dependent on the salt content and the temperature of the water. The effect of the salt content in fresh water is negligible, however, but the dependence on temperature is very great and of great biological importance because swimming and floating are affected by the viscosity. The reffect of temperature can be inferred from the values in Table 7. Water at 25°C thus has a viscosity which is only half that of water 0°C.

Table 7.

Temperature dependence of viscosity

Temp °C	Viscosity	
	Pa s x 10^3	%
0	1.787	100.0
5	1.561	84.8
10	1.306	78.7
15	1.138	63.7
18	1.053	58.9
20	1.002	56.0
25	0.890	49.8
30	0.798	44.7
35	0.719	40.3

ANNEXURE I

THE UNIQUE PROPERTIES OF WATER - A SERIAL

* Water, which is hydride of oxygen, is a unique chemical which has made life possible on earth.
* Life began in water and water still is the most essential ingradient for life.
* The uniqueness of water begins from the fact that at

normal temperatures this hydricde of oxygen is a liquid wheras all other hydrides of the elements neighbouring oxygen (ammonia, hydrogen chloride, hydrogen sulphide) are gaseous.

* Dielectric constant :

The electrostatic force of attraction between two oppositely charged particles in any medium is given by

$$F = \frac{q_1 q_2}{dr^2}$$

Where

F = force of attraction

q_1 = charge on one particle

q_2 = charge on the other particle

r = distance between the centres of the two particles

d = dielectric constant

Therefore, when other factors are equal, more the value of dielectric constant, less would be the value of F.

Water has a high value of dielectric constant - 80.1 at 20°C. As a consequence when any ionic substance is mixed with water, the force of attraction between the constituent ions of the substance, F, gets weakened due to the high value of water's dielectric constant. In other words water discourages ions to unite thus helping ionisation and dissolution.

This ability is so pronounced that water has earned the title of 'universal solvent'.

* Water also has other abnormal chemical and physical properties in terms of :

  Molecular structure density viscosity dielectric constant heats of fusion/vaporisation thermal conductivity surface tension

* All these emanate from the angular nature of water molecule :

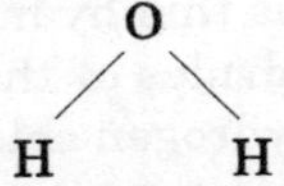

angle : 104.5° instead of 180°

which leads to hydrogen bonding and cluster formation Water molecules get bonded through hydrogen bonds :

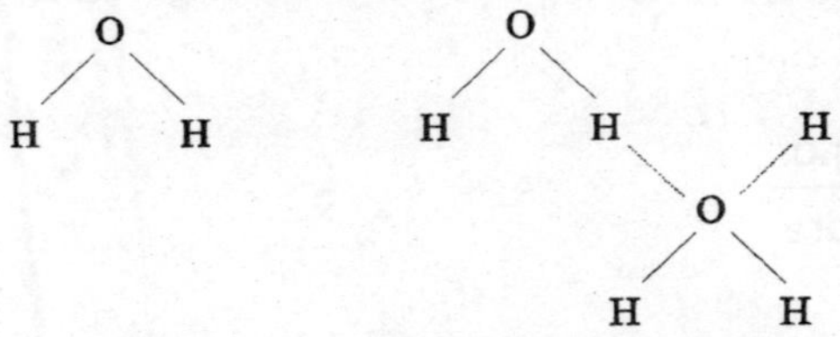

Which in turn leads to cluster formation (roughly tetrahedral arrangement; the bonds form and break within 10^{-11} seconds).

These dynamic 'liquid crystals' achieve such a geometry at near ~ 4°C that water gets densest at that temperature.

What makes tropical lakes more sensitive or 'fragile' compared to other lakes? The reasòns :

* Because of the peculier density - temperature relationships in water thermal stratification in these

  lakes is more pronounced. As a consequence chemical stratification is also very pronounced.

* Due to higher temperatures, the DO levels are lower - wheras a lake in a temperate region has ~ 11 ppm DO (at 10°C) in the epilimnion(i,e the upper layer of the lake upto which light penetrates), a lake in a tropical region (35°C) would have only 60% of this level.

* Due to higher temperatures, microorganisms are more active. Therefore for the same concentration of organic matter, BOD load on tropical lakes is higher.

* Tropical sun-shine adds primary production; it also hastens decay of dead plants and animals - these further contribute to pollution stress through BOD.

* Higher hypolimnic temperatures and lower hypolimnic DO levels combine to facilitate luxary release of nutri-

ents from the sediments. Both these factors aid eutrophication.

* In general tropics are thickly populated; there is, therefore greater demand on water and greater habitation - related pollution of the lakes.

Strategies for rehabilitating and managing tropic lakes

* Cut-off pollutants
  . by strategies such as using alternative detergents
  . by trating run-in water
    - point
    - non-point
* Imporove the lake condition
  . hypolimnic aeration
  . drawing off hypolimnic water
  . removal of excess
  . dredging, deweeding, aquaculture
* Improve the lake catchment
  . afforestation, turfing
  . regulating fertiliser and pesticide use.pa

ANNEXURE II

BUNDS OF BOOM AND DOOM

In India, during 1951-85, 17 million hectares of irrigation potential was developed with large dams, but -

* Over a million ha of forest was lost
* Waterlogged areas increased to 6 million ha (35% of irrigation area)
* Areas under salinity increased to 7.05 million ha (41.5% of irrigated area)
* Groundwater table in many command areas rose alarmingly : in Bhakra it has been rising at the rate of 1 m/yr

* In humid and sub-humid areas introduction of canals actually led to reduction in crop productivity
* Soil erosion increased : in Kali project (Karnataka), for example loss of top-soil in catchment area was severe enough to make it a desert unfit for cultivation for all times
* Weeds grew explosively ...
* Build-up of pollution occured : many of the damed rivers became dull and dirty, like Beas in Punjab
* There was adverse impact on public health, tribal habitats, fisheries
* The benefits in arid areas are spectacular for the first 10-20 years but the rot then sets in with gradual increase in waterlogging and salinity
* Floods were supposed to decrease, but actually :

Year	*Cropped area affected by floods (million hactares)*	*Total damage at current prices*	*(million rupees) at 1952-53 prices*
1953	0.93	540	520
1959	1.54	790	680
1968	2.69	2030	950
1971	6.24	6320	2630
1974	3.30	5690	1400
1976	7.68	8890	2390
1977	8.25	12000	2820
1978	10.05	14550	3420
1980	5.41	8300	1950

IN ULTIMATE ANALYSIS :

The average national yield for irrigated land is only 1.7 tonnes of grain per ha against the target of 4-5 tonnes. And the reservoirs are silting so fast they would soon have no space left for water :

Reservoir	*Assumed rate of siltation (acre/year)*	*Actual rate of siltation (acre/year)*
Bhakra (India)	23,000	33,475
Maithon (India)	684	5,980
Mavurakshi (India)	538	2,000
Nizamsagar (India)	530	8,725
Panchet (India)	1,982	9,533
Ramganga (India)	1,089	4,366
Tungabhadra (India)	9,796	41,058
Ukai (India)	7,448	21,758
Bridge Port (USR)	-	818
Kalabagh (Pakistan)	-	1,60,000
Tarbela (Pakistan)	-	1,29,000

14

ENVIRONMENTALLY BENIGN GENERATION OF ENERGY FROM AQUATIC BIOMASS

In many regions of the world, the production of aquatic biomass is so high that biomass-based fuels can more than fulfill the energy needs of those regions. Amongst the fastest growing aquatic biomass, and therefore amongst the most attractive sources of renewable energy, are aquatic weeds. But, as described in Chapter 6, the weeds cause enormous harm to the water resources, in terms of quantity as well as quality. The weeds also cause major hindrance in agricultural production. By utilising the weeds as energy source these negative impacts can be prevented to a great extent while generating clean and valuable energy.

In this chapter an attempt has been made to describe aquatic biomass in the perspective of the overall biomass resources of the world. The growth potential of the aquatic biomass and its impact on the environment in the context of utilisation as a management option has been briefly reviewed. Anaerobic digestion has been discussed as the technology most appropriate for *producing energy from aquatic biomass. In the concluding portion of the chapter, salient features of research and development efforts vested so far in utilising aquatic biomass are presented. The successes achieved, the major problems remaining to be solved and the global efforts in addressing these problems have been highlighted.*

BIOMASS AND AQUATIC PLANT BIOMASS

Biomass is the general term used to include 'phytomass' or plant biomass and 'zoomass' or animal biomass. Sun's energy when intercepted by plants and converted by the process of photosynthesis into chemical energy, is 'fixed' or stored in the form of terrestrial and aquatic vegetation (Figure 1). The vegetation when 'grazed' (used as food) by animals gets converted into 'zoomass' (animal biomass) and excreta. The excreta from terrestrial animals, especially dairy animals, can be used as a source of energy while the excreta from aquatic animals gets dispersed and wasted as it is not possible to collect it and process it for energy production. The upper limit of capture efficiency of solar radiation in biomass may be as high as 15% but in most of the species it is generally 1% or lower (1). The energy thus captured by photosynthesis is a small percentage of the total solar energy reaching our planet, but the total volume of biomass created is still very large compared with our energy needs. The dry weight of all living phytomass on the earth's land surface has been estimated as nearly $17x10^{11}$ tonnes, produced at a rate of almost 10^{11} tonnes per year (2). Nearly all of this, together with undecomposed dead biomass in forest and field, as well as large quantities of aquatic vegetation, are potentially capable of being converted into some form of energy.

The value of the world's plant biomass resource can be gauged from the fact that in many regions of the world, fuels derived from plant biomass can more than satisfy the energy needs of those regions. For example, it has been estimated that Sudan and Brazil can potentially obtain six times and two times their current energy needs respectively from their biomass resources (3).

In comparison the animal biomass, or 'zoomass' represents a much smaller potential for conversion into energy. Indeed, apart from work performed by animals and humans whose 'fuel' is feed and foodstuff, and from rare products such as fish oil, whale oil and tallow, most of the animal biomass used or likely to be used for energy production as in the form of excreta.

In countries like China and India, where per capita energy consumption is low and the population of dairy animals is

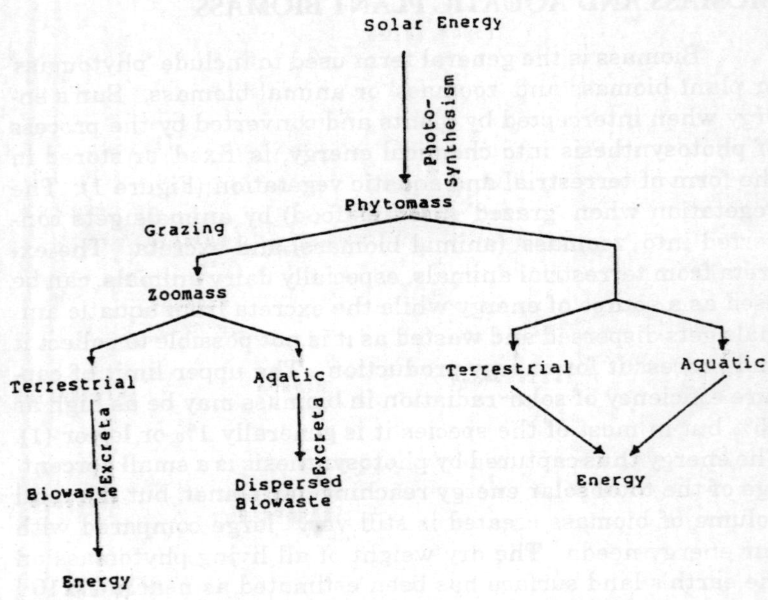

Figure 1: The Solar Energy-Biomass Energy Pathways

large, even the excreta may provide a sizable fraction of the total energy requirement but in general animal biomass contributes very little to the overall biomass potential of the world, while the contribution of aquatic animals to the overall aquatic biomass potential is virtually insignificant. Therefore subsequent discussion will focus on aquatic plant biomass and the term 'biomass' will be used to denote only the phytomass.

Of the overall biomass growing in the world, as much as 33% comes from aquatic vegetation. In other words, aquatic vegetation contributes approximately 1/3 as much to the world's primary production as terrestrial plants do (4). How much of this is contributed by aquatic weeds like water hyacinth (Eichhornia crassipes, Mart, Solms), Salvinia (Salvinia molesta,

Mitchell), hydrilla (Hydrilla verticillata) and others, has not yet been estimated but the contribution is likely to be immense considering the fact that some of these species produce more biomass per unit area than even the most prolific of their terrestrial counterparts (Table 1). Higher aquatic plants are not generally water-limited unlike the majority of land plants, and are capable of higher rates of photosynthesis by keeping their stomata open longer than land plants, thereby increasing CO_2 absorption and net solar energy gain. The ease of reproduction through vegetative propogation, high tolerance towards environmental fluctuations, capability to grow on bad quality as well as good quality waters - all these factors combine to make aquatic weeds like water hyacinth and salvinia forceful colonisers of water bodies (10). In fact many authors consider the growth potential of water hyacinth an salvinia as the ultimate achievable by any plant, terrestrial or aquatic, on earth (11). Similarly certain species of microalgae, such as spirulina (Spirulina sp) and macroalgae such as the marine brown kelp (Macrocystis pyrifera) or red seaweed (Gracilaria tikvahiae) have productivities higher than most of the terrestrial plants.

THE NUISANCE OF THE AQUATIC WEEDS

The high productivity of aquatic weeds can be an asset only if ways and means may be found to utilise them profitably. Otherwise the weeds are a very major nuisance (10,12). They cause immense harm to the water resources by occupying lakes, ponds and canals and thus reducing the carrying capacity of these water bodies. The evapotranspiration by the weeds causes water losses several times higher than the losses due to evaporation from weed-free surfaces. The weeds compete with useful crops like rice for nutrients, sunlight and space. They interfere with fisheries and inland navigation. They also harm the water quality in several ways : cutting off sunlight from reaching the water; reducing the dissolved oxygen levels of the water; polluting the water through decay; causing stagnation and thereby supporting growth of mosquitoes and flies; impeding the growth of useful aquatic organisms such as fishes and prawns ... and so on. The combined impact of all these factors can be strong enough to cause tremendous hardships to people and jeopardise the entire economy of the affected region. Attempts to destroy the weeds with chemicals or bioagents (insects, snails, ducks, fishes, rats etc) have failed throughout the world (12). At best such attempts only bring temporary relief because some or the other weed soon replaces the destroyed one. Worst still, introduction of chemicals or bioagents on a large scale carries the grave risk of ecological damage.

Table 1.

Biomass growth potential of selected aquatic and terrestrial plants

Plant	*Productivity (tonnes ha^{-1} yr^{-1})*	*Reference*
Aquatic		
Water hyacinth	62-104	5
Salvinia	54-102	7
Water pennywort	8-41	6
Spirulina	79-33	8
Giant brown kelp	44-54	8
Red seaweed	32-60	9
Terrestrial		
Sugar cane	45-50	4
Bermuda grass	7-11	4
Eucalyptus	20-48	4

UTILISATION OF AQUATIC WEEDS

Attempts have been made to utilize (12,13) aquatic weeds as food (duckweed, cattails) animal feed supplement (water hyacinth, duck-weed, salvinia), raw materials for paper pulp (water hyacinth, salvinia), source of commercially useful chemicals (hydrilla, water lily), and compost (duckweed, water hyacinth, salvinia). These attempts, reviewed recently (10-12), have met with very limited success either because they enable only a very small fraction of the weeds to be utilised (for example as compost or as animal feed supplement) or because they fail to compete with the existing alternatives (for example as raw materials for paper pulp). The only method which holds promise of utilising aquatic weeds on a large scale, compatiable with their high growth potential, is convertion to energy.

CONVERTION OF BIOMASS INTO ENERGY

Several highly developed technologies are available for converting biomass into energy (Table 2). Of these options, thermal and thermochemical convertion are very widely used for obtaining energy from terrestrial biomass, especially woody material. Amongst fermentation-based technologies, aerobic fermentation is used on a large scale for obtaining alcohols from sugarcane juice and other high-carbohydrate agricultural products. However, these options are unsuitable for aquatic biomass mainly because the water content of the aquatics is very high - 90 to 95%. The two most prolific aquatics : water hyacinth and salvinia, contain over 92% water. Drying these plants to a level at which they can be used in thermal and thermochemical process is too costly to be practicable. Likewise the sugar content of the aquatics is too low to make them attractive as raw material in alcohol production. The most appropriate and feasible process for energy production from aquatic biomass is anaerobic digestion (4).

ANAEROBIC DIGESTION OF AQUATIC BIOMASS

Anaerobic digestion is bacterial fermentation of organic wastes in the absence of free oxygen. The fermentation, when carried out for producing energy, leads to the breakdown of complex biodegradable organics in a four-stage process (Figure 2). If the process is properly controlled so that it proceeds as per these stages the principal end product is methane gas, containing about 35% carbon dioxide and traces of ammonia, hydrogen sulfide and hydrogen. This end product, commonly called 'biogas', is a very convenient and clean fuel and can either be used directly, with or without the removal of carbon dioxide, or can be converted into electricity with the help of suitable generators. An aquatic plant can yield about 35-40 litre biogas/kg of plant (fresh weight) (7,14). At conservative estimates, fast growing aquatics like water hyacinth and salvinia attain annual productivities of 60 tonnes ha^{-1} yr^{-1} (dry weight basis) (5,7) equivalent to 800 tonnes ha^{-1} yr^{-1} on wet weight basis. On anaerobic digestion this biomass would yield close to 30,000 m^3 of biomass ha^{-1} yr^{-1}, equivalent to 225 million K cal.

Considering the immense quantities of the aquatic biomass crops already standing, the little or no commercial market for them which may compete with their use in energy production, and the vast ocean surfaces available for the further cultivation of marine biomass, one would assume that aquatic biomass should be one of the most desirable raw materials for energy production. But this is not so due to certain technological difficulties. These difficulties are daunting but are not unsurmountable.

Table 2.
Technologies for converting biomass into energy

A.	THERMAL CONVERSION	
	i.	Direct Combustion
	ii.	Pyrolysis
	iii.	Gasification
B.	THERMOCHEMICAL CONVERSION	
	i.	Conversion to methanol
C.	FERMENTATION	
	i.	Aerobic fermentation
	ii.	Anaerobic digestion
D.	UNPROVEN NOVEL CONCEPTS	
	i.	Biophotolysis
	ii.	Extraction of hydrocarbons
	iii.	Hydrogasification
	iv.	Fuel cells

PROBLEMS AND EMERGING SOLUTIONS

Till the 1970s, anaerobic digestion was mainly used to stabilise sewage sludge, dairy manure or high-strength (in terms of COD & BOD) biowastes such as distillary spentwash from industries. When aquatic biomass began to be used in digesters, it was found to pose some special problems. Firstly it needs to be shredded or minced before feeding to the anaerobic digesters so that it can be easily attacked by bacteria. Secondly it is lighter than water, unlike sewage sludge or manure, with the result that it floats to the top of the water column in the digesters, forming a thick layer of scum and impeding the digestion process. Thirdly it does not easily homogenise with water and is difficult to feed in or remove from the digesters in a continuous process. For these reasons the millions of small 'family size' digesters so common in Asia, Africa and Latin Amercia or the medium and large size 'farm digesters' common in Europe, Canada and USA are not suitable for digesting aquatic biomass. If aquatic biomass is fed to such digesters in place of farm manure, the digesters clog sooner or later and cease to function efficiently (14).

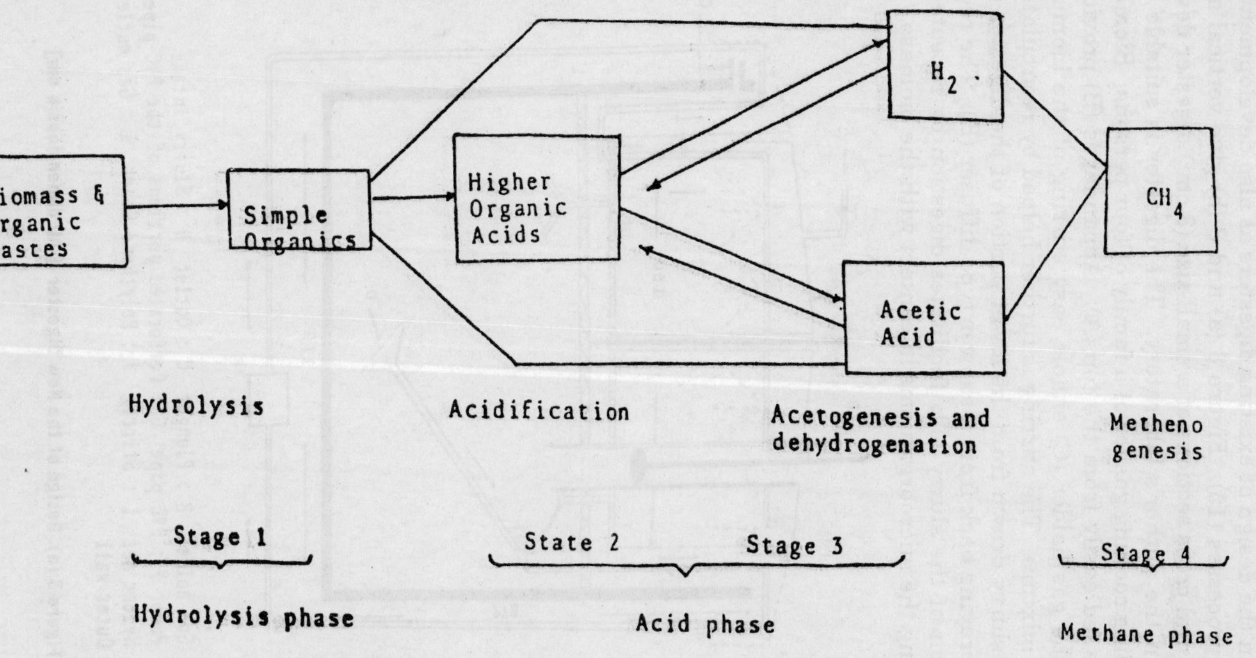

Figure 2. Anaerobic digestion: the three main phases involved

Attempts to solve these problems have led to modifications in the design of existing digesters as also development of novel processes (15). Figure 3 (a) and 3 (b) give vertical and horizontal cross sections of a small-size (3 cm^3) digester developed in the author's laboratory. The digester is suitable for providing cooking gas for of a family of four persons. Biomass can be fed easily from the top (A). Appandages (B) provided with the gas holder (C) enable easy stirring of the biomass-water mixture. The stirring is further helped by recirculation of the slurry drawn from the lower portion of the digester (D) and spraying back from a shower type diffuser (E). The recirculation of the slurry also facilitates digestion by repeatedly bringing the microorganisms in contact with the biomass.

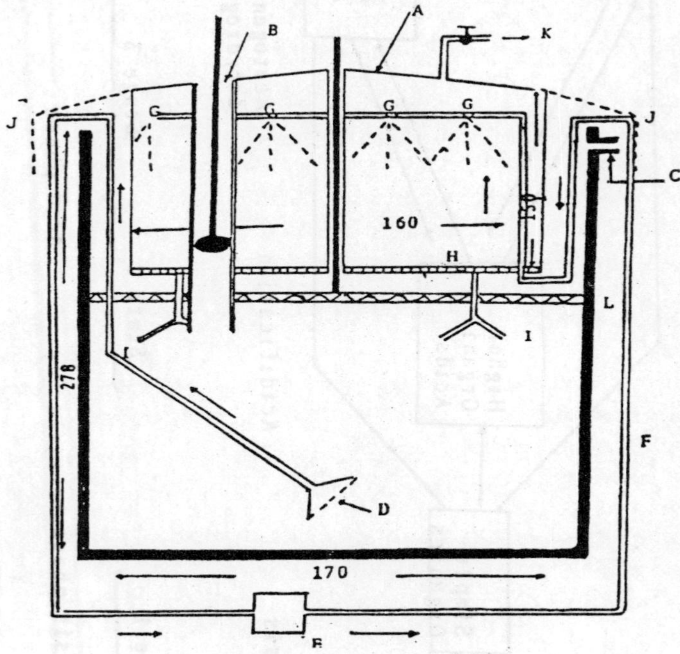

A : Gas holder B : Plunger C : Outlet D : Slurry Intake
E : Pump F : PVC pipe G : Perforated portions of the PVC pipe
H : Nylone net I : Stirrer J : Polythene Cover K : Gas outlet
L : Outer wall

Figure 3 (a). Design of the New Digester [All dimensionsin cm]

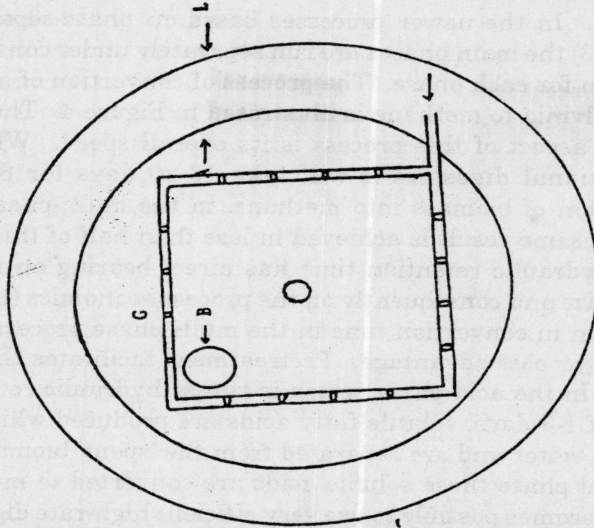

Figure 3 (b). Horizontal section showing the position of the plunger and slurry shower; the polythene sheet (not shown here) covers the space between A & L

One approach which holds great promise is solving the technical problems associated with the anaerobic digestion of aquatic biomass, is 'phase-separation' or 'multi-stage digestion'. As is evident from Figure 2 anaerobic digestion is essentially a three-phase process. In the first phase cellulolytic microorganisms convert complex organic matter (cellulose, hemicellulose) into simpler organics. In the second phase which may be called 'acid phase' and which includes the 'acidification stage' and the 'acetogenesis and dehydrogenation stage', the acidogenic bacteria convert the organics to higher fatty acids (propionic acid, butyric acid etc) and then acitogenic bacteria convert these acids to acetic acid and hydrogen. Finally in the third and last phase, which is termed 'methane phase' the methanogenic bacteria produce methane.

Kinetically the first and second phases are faster than the third phase. Also while the first and second phases do not require rigid control of pH or temperature, the methane phase does. The methanogenic bacteria are much more sensitive to the presence of inhibitors and toxicants than the cellulolytic or acid-phase bacteria. In the convertional anerobic digestion processes all the three phases are made to carry out together in a single tank, with the result that proper kinetic control is not possible. In the newer processes based on phase-separation (Figure 3) the main phases are run separately under conditions optimum for each phase. The process of convertion of aquatic weed salvinia to methane is illustrated in Figure 4. The most notable aspect of this process is its overall speed. While in conventional digesters it will take 15-20 days for 60-70% convertion of biomass into methane, in the multi-phase process the same result is achieved in less than half of this time. Since hydraulic retention time has direct bearing on the digester size and consequently on the process economics (16), the reduction in convertion time in the multi-phase process leads to a major cost advantage. Pretreatment facilitates the first phase. In the acid phase which is run at hydraulic retention times of 1-3 days, volatile fatty acids are produced which dissolve in water and are separated from the 'spent' biomass. In the final phase these soluble acids are converted to methane and it becomes possible to use very efficient high-rate digesters like anaerobic filter, upflow anaerobic sludge blanket reactor, or expanded bed/fluidised bed reactors. Such digesters can accept only soluble feed and without phase separation and consequent solubilisation of the reactants, the biomass can not be converted into energy using such digesters.

Other promising approach is 'dry digestion' or digestion of partially dried biomass with little or no homogenising with water (17). This may be an attractive technology if process control can be achieved.

Amongst novel unproven technologies for converting aquatic biomass into energy (Table 2, D) mention may be made of the culture of green hydrocarbon-producing microalgae *Botryococcus braunii* (18). The algae have a hydrocarbon content of 20-52%. Earlier, microalgae have been investigated as a source of liquid fuels in the form of vegetable-type oils (lipids) for over 40 years but have not been found cost-effective.

In recent years, aquatic weeds have been successfully employed for wastewater treatment on a large scale in several countries (19, 20). The weeds growing on wastewaters need to be periodically harvested and disposed off. Successes in developing cost-effective methods of converting such harvests into energy are expected to provide a major incentive for the adoption of aquatic plant based waste treatment processes all over the world.

Figure 4. Anaerobic digestion: of aquatic weed Salvinia by three-phase process

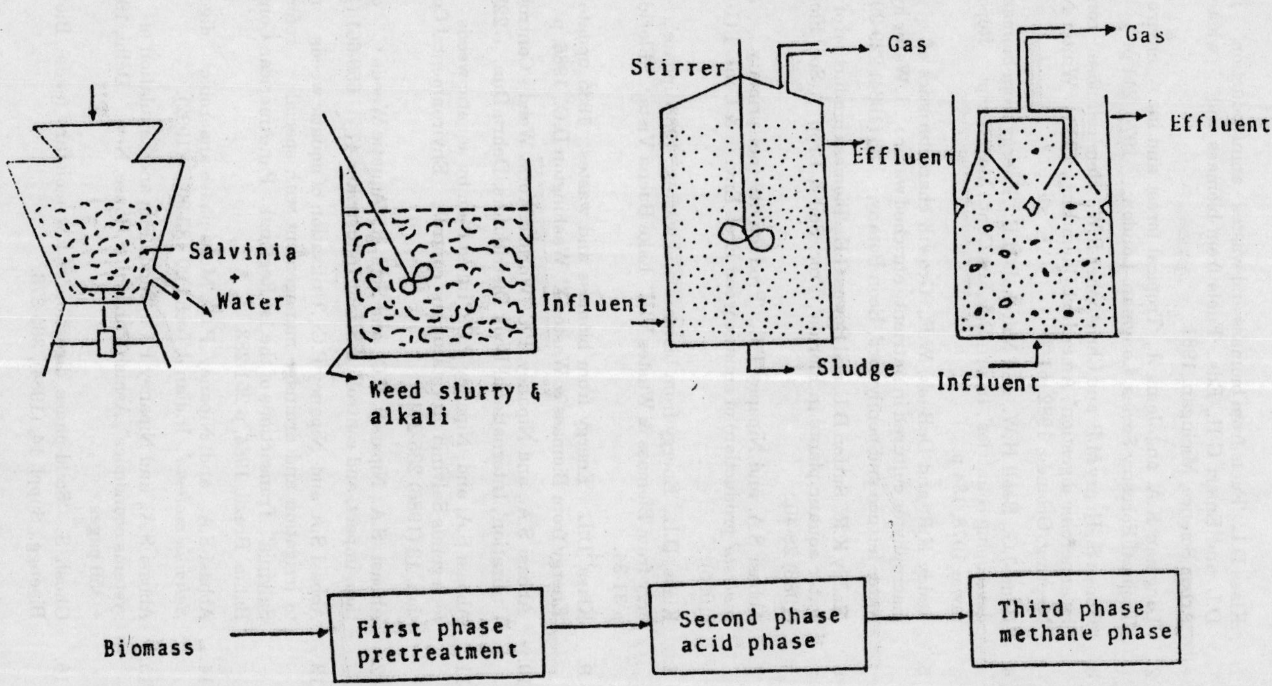

REFERENCES

1. Klass D.L., 'Fuels from biomass and wastes - anintroduction' in Klass D.L. and Emert G.H., Eds., 'Fuels from biomass and wastes', Ann Arbor Science, Michigan, 1981.

2. Longman K.A. and Jenik J., 'Tropical forest and its environment', Tropical Forestry Series, Longman, London; 1974, 211 p.

3. Ghosh S., Henry M.P. and Christopher R.W., 'Hemicellulose convertion by anaerobic digestion', Amer.Inst.Chem.Eng., 1982 Winter National Meeting, Orlando, 1982, p1.

4. Bene J.G., Beall H.W., and Marshall H.B., 'Energy from biomass for developing countries', Intern.Dev.Res.Cent., Manuscript Reports, Ottawa, 1978, 134 p.

5. Reddy K.R. and DeBusk, W.F., 'Growth characteristics of aquatic macrophytes cultured in nutrient - enriched water : I. Water hyacinth, water lettuce and pennywort', Econ.Botany, 38 (1984) 229-39.

6. Reddy K.R., Sutton D.L. and Bowes G.E., 'Biomass production of freshwater aquatic plants in Florida', Proc. Soil & Crop Soc. Florida, 42 (1983) 28-40.

7. Abbasi S.A. and Nipaney P.C., 'Productivity (net primary *Salvinia molesta*' production) of aquatic weed Ecol. Env. & Cons. 1 (1-4) 11-22 (1995).

8. Klass, D.L., 'Energy from biomass and wastes : 1983 update', in Energy from Biomass & Wastes VIII', Lake Buena Vista, Florida, 1984, p 31-34.

9. Klass, D.L. 'Energy from biomass and wastes : 1985 update', in 'Energy from Biomass & Wastes X', Washington D.C., 1986, p 50.

10. Abbasi S.A., and Nipaney, P.C. 'World's Worst Weed - Control and Utilization', International Book Distributors, Dehra Dun, 228 pages.

11. Abbasi S.A. and Nipaney P.C., 'Infestation by aquatic weeds of the fern genus Salvinia : Its status and control', Environmental Conservation, 13 (1986) 235-241.

12. Abbasi S.A., Nipaney, P.C., and Soni, R. "Aquatic Weeds : distribution, impact, and control", J.Scient.Industr.Res.47 650-661 (1988).

13. Abbasi S.A. and Nipaney P.C., 'Utilisation of aquatic weeds relevant to irrigation and drainage management with special reference to Salvinia'. Transactions of the 1st Regional Pan-American Conference, Bahia, Brazil, 1984, p 251-283.

14. Abbasi S.A., and Nipaney P.C., 'Multi-phase anaerobic digestion of *Salvinia molesta*'. Indian J.Tech., 30 483-90 (1992).

15. Abbasi S.A., and Nipaney P.C., "Modelling and simulation of biogas systems economics", Ashish Publishing House, New Delhi, 1993; xviii + 356 pages.

16. Ghosh S, 'Soild-phase digestion of low moisture feeds', Biotechnol. Bioengg., Suppl. 14, (1984), 365-376.

17. Metzger, P., 'Screening of wild strains of the hydrocarbon- rich alga *Botryococcus braunii*, productivity and hydrocarbon nature', in Palz W., Coombs J, and 'Hall D.O., Eds., 'Energy from biomass : 3rd E.C. Conference', Elsevier, London, 1985, p 727-31.

18. Abbasi S.A., 'Aquatic plant based water treatment sysems in Asia', in K.R. Reddy and W.Smith, Eds., 'Aquatic plants for water treatment and resource recovery', Proceedings of the International Conference, Orlando (July 1986), in press, 31 p.

19. Reddy K.R. and DeBusk W.F., 'Nutrient removal potential of selected aquatic macrophytes', J.Environ.Qual., 14 (1985) 459-462.

15

TRACE METALS IN FRESHWATER ECOSYSTEMS

In Chapter 2 I have narrated the Minimata story. It revolved round a trace metal-mercury. There are numerous other horror stories with trace metals as their focus-Itai Itai involving cadmium, Genu Velgam involving molybdenum.....and so on. What are trace metals?

WHAT ARE TRACE METALS?

Trace metals are elements present in low quantities in nature. Several of them are essential for living organisms eventhough they are needed for this purpose in very small amounts. The normal concentration of all trace elements combined in living organisms is less than 2% of the dry weight of the organisms.

Those of the trace metals which are essential for plant or animal life play vital role in the organism's biochemical and metabolic activities. But at higher - than - essential levels the metals can prove toxic. Thus maintenance of a delicate balance between the biologically available and unavailable forms of various trace metals is crucial for the existence of a healthy eco-system.

Natural waters and associated particulate matter consist of numerous elements including heavy metals, inorganic compounds, and organic compounds. These species are in a continual state of interaction which is governed by intricate and dynamic sets of physiochemical and biological processes (like annual, seasonal and daily variations in pH, temperature, dissolved oxygen, chelating capacity of water, uptake and excretion of chemicals *via* food etc).

NATURAL SOURCES OF METALS IN THE ENVIRONMENT

In terms of geological history, chemical weathering and volcanic activities have been the major release mechanisms responsible for the trace metal concentrations in the environment. Chemical weathering of igneous and metamorphic rocks in drainage basins contributes the background levels of trace metals entering surface waters. Decomposing plants and animal detritus also contribute small, yet significant, amounts of metals to surface waters. Precipitation and atmospheric fallout are the second-most important sources of trace metals in freshwaters. Metals in the atmosphere are derived from :

i) dust from volcanic activities;

ii) erosion and weathering of rocks and soils;

iii) smoke from forest fires; and

iv) aerosols and particulates from the ocean surface.

In addition, human activities like mining operations, domestic sewage discharge, agricultural operations and industrial wastewater discharge, are major sources of release of metals into environment.

FORMS OF OCCURRENCE OF METALS

In order to study the behaviour and fate of trace metals in natural waters, we must first look at the various forms that they can occur in and how these are formed.

Trace metals can occur in three physical forms : suspended (>100 um), colloidal (1 um to 100 um) and soluble (<1 um). Further, they occur in numerous *chemical* forms of the physical types mentioned above (Figure 1).

HOW THESE SPECIES ARE FORMED?

Metals in natural waters may exist simply in the form of free metal ions surrounded by coordinated water molecules, although concentration of anionic species (OH^-, Cl^-, SO_4^{2-}, HCO_3^{2-}, amino acids) may be sufficient to form inorganic or organic complexes with hydrated metal ions by replacing the water molecules. Other types of associations also occur with colloidal and particulate material like clays, hydrous iron, manganese oxides and organic debris. The specific reaction mechanisms are as follows:

1) *Formation of inorganic ion pairs* : Most highly-charged metal ions are strongly hydrolyzed in aqueous solution which produces charged ion pairs, insoluble metal oxides or hydroxides.

2) *Formation of complexes*: Metal ions react with inorganic ions HCO_3^-, F^-, PO_4^{3-} and S^{2-} to form soluble complex ions; the extent of complexation depends on the metal concentration, the ligand concentration and pH of the water.

3) *Formation of organometallic compounds*: Metals can also bind with natural and synthetic organic substances yielding organometallic compounds such as salts of organic acids and metal-organic complexes.

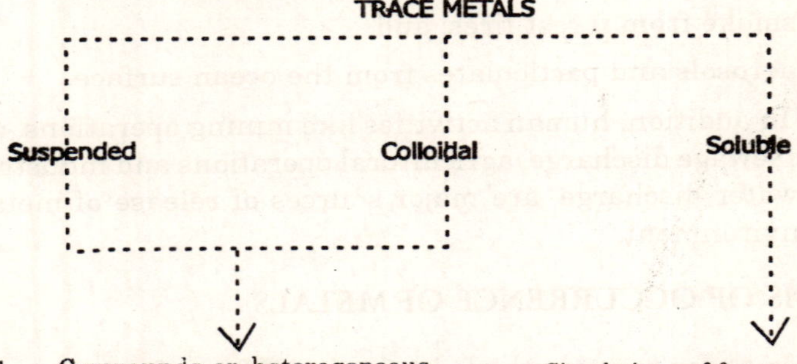

1. Compounds or heterogeneous mixtures of metals in forms like hydroxices, silicates, oxides and sulphides which often form highly dispersed colloids.
2. Organic matter to which metals are bound species sorbed on colloids like clay.
3. Metal species sorbed on colloids like clay.
4. Precipitates, organic particles and remains of living organisms.

1. Simple ions of free metals.
2. Complex ions (inorganic ion pairs).
3. Unionized organo metallic chelates or complexes.
4. Inorganic complexes.

Figure 1. Form of occurrence of metals on limnic environments

The concentration of dissolved metals found in the limnetic zones of lakes depends on the interplay between uptake by phytoplankton, release by decay from sediments, and input, from rivers. More specifically it depends on pH, type and concentration of ligands (including chelating agents) and oxidation state of the metal. In the pH range of natural waters (between 5 and 9.5) and in aerobic conditions free metal ions occur mainly at low pH. With increasing pH, carbonates, oxides, hydroxides and silicates precipitate out. These interactions are discussed in detail in the next section.

The interactions between the aqueous and solid phases of trace metals (Figure 2) basically involve two processes :

1) The formation of metal substrates, and deposition and enrichment of trace metals in the sedimentary environments, and 2) the mobilisation of metals from sediments.

The formation of metal substrates occurs in the following ways :-

a) precipitation as hydroxides, sulphides and carbonates as a result of exceeding solubility product at pH>7-8 and presence of free oxygen. For example, Fe^{2+} iron precipitates out as $Fe(OH)_3$ when the waters containing this ion are oxygenated;

b) absorption and co-precipitation of cations and anions from the aqueous phase by hydrous oxides of Fe,Mn and Al;

c) coagulation, flocculation of soluble and colloidal material, direct precipitation, and adsorption into sedimentary materials lead to incorporation of metal-organic species into or onto sediments;

d) carbonate adsorption and co-precipitation of trace metals - for example for Zn,Co,Cd and Pb in alkaline environments. The following factors affect the formation of metal substrates:-

a) presence of heavy minerals (having sorption capacity for heavy metals) in silt and sand;

b) presence of organic matter, hydrous iron and manganese oxides which act as effective sinks for metals;

c) presence of inorganic ions : for example phosphate is an important precipitating agent for iron through such process as adsorption of PO_4^{3-} on ferric hydroxide;

d) pH level : it is an important determinant of the ratio of dissolved to undissolved metals in limnic environments - for example, ionic copper is rapidly lost by precipitation as malachite ($CuCO_3$) when pH levels fall between 6 to 8.5.

The mobilization of trace metals from sediments takes place by way of the following major processes :

a) at elevated salt concentrations, the alkali and the alkaline earth cations compete for adsorption sites on the solid particles and colloids, thereby displacing the sorbed trace metal ions;

b) changes in redox conditions such as a decrease in oxygen potential in sediments (due to advanced eutrophication) results in changes in chemical form of the metals and consequently a change in water solubility. For example, when lake sediments enriched with Fe^{3+} become anoxic in summer, the Fe^{3+} - Fe^{2+} equilibrium shifts to the right releasing soluble ferrous ions;

c) reduction of pH leads to dissolution of carbonates and hydroxides and increased desorption of metallic cations due to competition with hydrogen ions;

d) increased use of natural and synthetic complexing agents can form soluble stable metal complexes with trace metals that are otherwise adsorbed onto solid particles;

e) biochemical transformations can lead to transfer of metals from sediments into the aqueous phase or their uptake by aquatic organisms and subsequent release via decomposition products.

SEASONAL INFLUENCES ON THE AQUATIC CHEMISTRY AND BIOLOGY OF METALS : THE SEASONAL CYCLE OF IRON SEASONAL CYCLE OF IRON

The chemistry and biology of trace metals follows seasonal cycle in most limnic ecosystems except in polymictic lakes. Discussed below is the illustrative example of iron (Figures 3 &

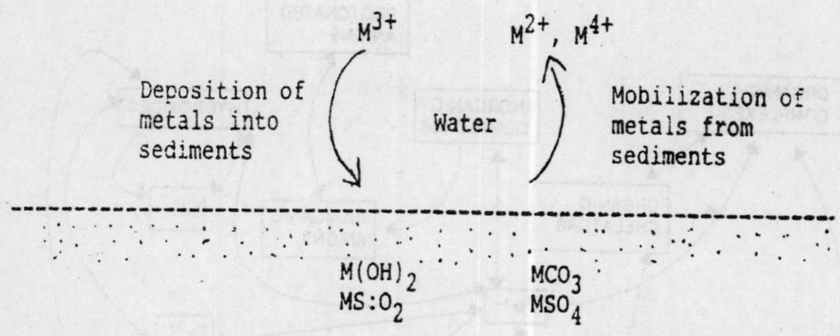

Figure 2. Interactions between aqueous and solid phases

4) which shows a seasonal behaviour in the hypolimerion.

In well-oxygenated waters, iron occurs but is rare because the insolubility of its trivalant form Fe^{3+}. During the spring overturn most of this metal is in the sediments as $Fe(OH)_3$, $FePO_4$ and $Fe_2(CO_3)_3$. An oxidised microzone of iron-containing molecules in a complex colloidal layer seals nutrients within the sediments and little escapes to the overlying water. That is how the matter rests throughout the summer stagnation period in a stratified oligotrophic lake.

During spring overturn, the disappearance of the oxidised barrier and release of nutrients to the supernatant water leads to the consumption of more oxygen and perhaps to the escape of more nutrients, consequently to a rapid conversion of Fe^{2+} to Fe^{3+} (with immediate precipitation of $FePO_4$ and other iron compounds).

Thus the hypolimnion is similar to an iron trap - most of the iron that arrives there is retained, alternating between mobile-soluble and immobile-insoluble states.

ROLE OF AQUATIC BIOTA IN TRANSPORT AND TRANSFORMATION OF TRACE METALS

The lives of the aquatic biota and the trace metal cycling processes are intimately connected to each other in various ways. These processes can be studied by considering separately metal uptake, regulation and excretion processes.

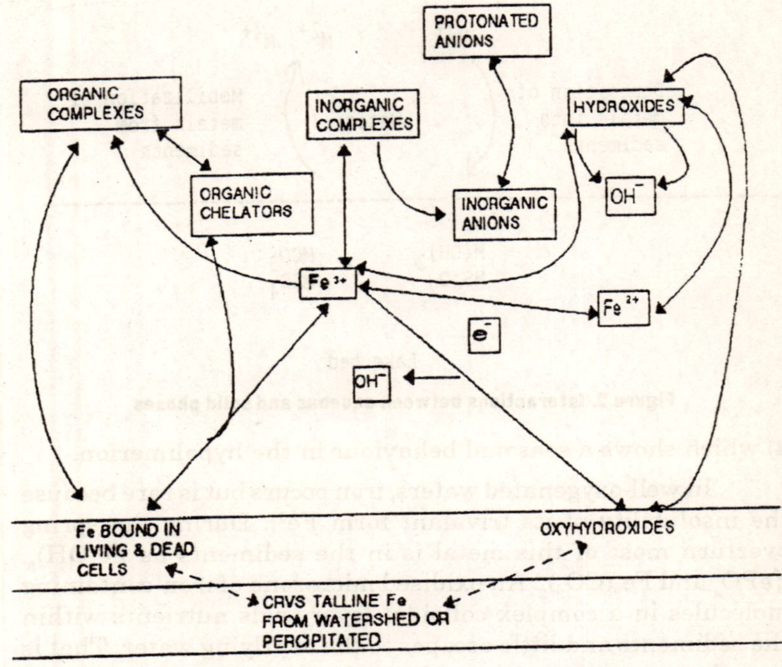

Figure 3. Iron cycle in limnic ecosystems

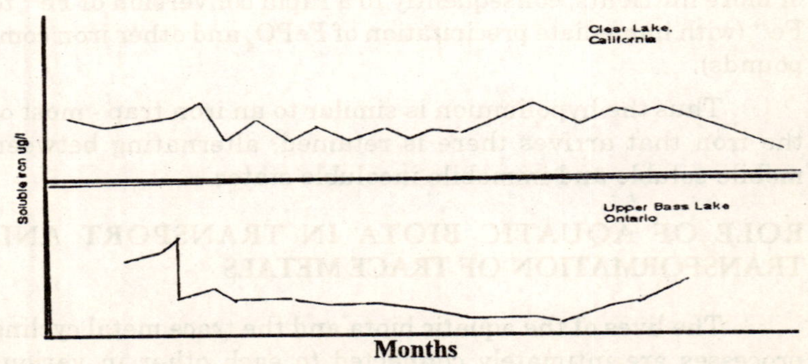

Figure 4. Seasonal cycles of dissolved iron in upper mixed layers of two lakes in California (USA) and Oritario (Canada)

UPTAKE PROCESSES

Metals can be taken up by aquatic organisms through the following main pathways :

1) from water through respiratory surfaces like gills by passive absorption; this pathway is mainly seen in fish;
2) adsorption from water onto body surfaces as in burrowing polychaetes;
3) uptake in ingested food or water through the digestive system as in crustaceans and molluscs;
4) directly from the solution in case of autotrophic organisms like macroalgae and phytoplankton which takes place through adsorption, diffusion, ion-exchange, etc through the cell wall or through absorption of metals from interstitial pore water *via* the root system.

Several factors affect metal entry into aquatic organisms:

a) the number of modes of metal entry: In heterotrophic organisms modes of metal entry are much greater than in autotrophs thus they tend to accumulate much more metal;
b) changes in physico-chemical factors like temperature, pH and salinity : these influence rates of absorption drastically;
c) physiological and behavioural characteristics oforganisms: these influence the amounts of metals being ingested - for example detritivores ingest much more metal than fishes of the limnetic zone as the substrata is usually rich in metals.
d) levels of availability of metals in the aquatic environment: these directly influence rates of absorption;
e) special properties of organisms : these enable them to absorb trace metals efficiently - for example blue-green algae gain a competitive advantage over other plankton for absorption of iron by extracellular secretion of powerful iron chelators called *siderochromes.*

REGULATION PROCESSES

Excretion of metals occurs through gills, guts, faeces and urine of organisms, and is the major process involved in the regulation of metal concentration in the animal blood. Some animals excrete high proportions of abnormal metal intake, in order to maintain a normal concentration. Even so, most organisms usually accumulate metals far in excess of their immediate needs. Non-essential metals such as mercury, cadmium, and lead are also accumulated. When the masses of accumulated metals in a prey are higher than the tolerance limit of the predator, metal poisoning occurs.

ROLE OF MICROBES

Microbes affect environmental transport of metals in three ways :

1) degradation of organic matter to lower molecular weight compounds which are more capable of complexing metal ions; thus microbes increase the quantum of biologically available metals;
2) alterations in the physicochemical properties of the environment and chemical form of metals by metabolic activities like changes in oxidation potential and pH conditions;
3) conversion of inorganic compounds into organometallic forms by means of oxidative and reductive processes, thus increasing biological availability of metals - for example bacterial methylation of mercury compounds into methyl mercury by methylating bacteria.

FOOD CHAIN TRANSFERS

Many but not all metals get substantially biomagnified in food chains - like mercury (Figure 5). The biomagnification process is characterized by the following features :

i) bioavailability of metals to animals at higher trophic levels is generally determined by transfer from water rather than prey organisms;
ii) filter-feeding organisms accumulate high levels of metals in their tissues but transfer only a small proportion to

predatory organisms;

iii) sediments and detritus usually contain highest metal concentrations in polluted limnic ecosystems and therefore sediment and detritus feeding organisms tend to accumulate higher metal concentration than animals at higher trophic levels;

iv) lifespan of animals at higher trophic levels is usually greater than organisms at lower levels; thus age-related enrichment may be a significant factor influencing the level of metal enrichment at the higher trophic levels;

v) different metals and their forms are taken up at different rates.

Figure 5

The factors effecting the toxicity of heavy metals are summarized in Figure 6.

TRACE METALS AND HUMAN ACTIVITIES

Influence of pollution on the global environment through activities like mining, burning of fossil fuels, agricultural practices and urbanization have accelerated release of trace metals and their salts in to the ecosphere. The present rate of global input of trace metals is found to be far in excess of the natural rates of their biogeochemical cycling.

The anthropogenic sources which bring metals into the environment are: contaminated wastes, deforestation of drainage basins, (leading to sediment flows into water bodies) and burning of fossil fuels.

Mining operations expose rocks to the atmosphere causing their accelerated weathering. For example, exposure of iron pyrites and other sulphide minerals to oxygen and moisture results in oxidation of the mineral and the formation of 'acidic mine drainage'. This causes serious water quality problems involving high levels of Fe, Mn, Zn etc.

In addition metal processing and refining operations can cause the diffusion and deposition of large quantities of metals into surrounding drainage basins or aquatic environments.

Appreciable amount of trace metals in domestic (effluents) is contributed by metabolic wastes, corrosion of pipes,

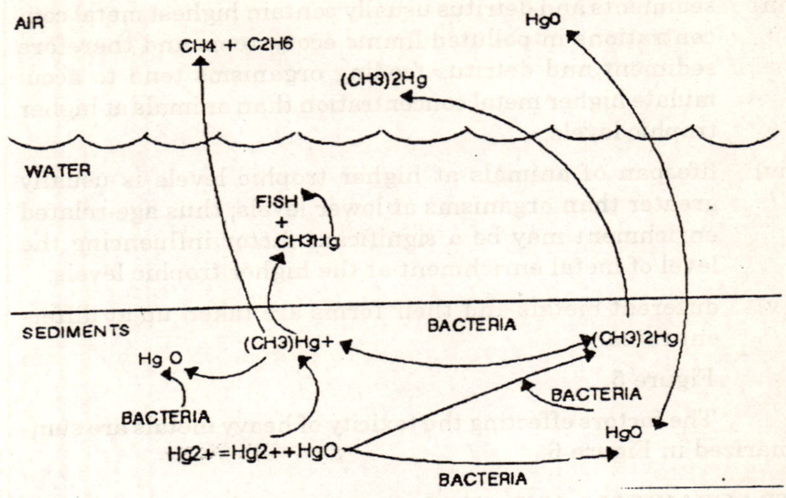

Figure 5. Biological cycle for mercury

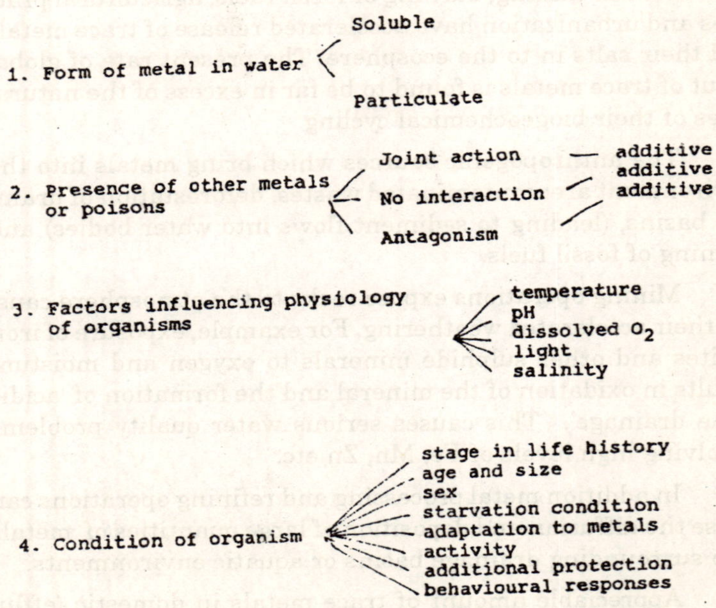

Figure 6. FActors affecting toxicity of heavy metals in solution

consumer products (like detergents) and sewage disposal. The metal composition of urban runoff is dependent on many factors like traffic, road construction and land use in the watershed.

Many trace metals are discharged into the aquatic environment through industrial effluents, dumping and leaching of industrial sludges and by sanitary (landfills).

Metal emissions from combustion of fossil fuels are a major source of airborne metal contaminants in natural waters.

Agricultural runoff containing trace metals from animal and plant residues, phosphatic fertilizers, specific herbicides and sewage nutrients also contribute to metal enrichment.

REFERENCES

Schwoerbel,J. (1987). Handbook of Limnology. Ellis Horwood Ltd, West sussese (5th edition).

Counell D.W., Miller G.J. (1984). Chemistry and Ecotoxicology of Pollution. John Wiley & Sons, Canada.

Goldman,C.R, Home, A.J. (1983). Limnology. McGraw-Hill Book Company, Japan.

Cole, G.A. (1979). Textbook of Limnology. 2nd edition. C.V.Masby Company, London.

PART -III

MEMOIRS

consumer products (like detergents) and sewage disposal. The metal composition of urban runoff is dependent on many factors like traffic, road construction and land use in the watershed.

Many trace metals are discharged into the aquatic environment through industrial effluents, dumping and leaching of industrial sludges and by sanitary landfills.

Metal emissions from combustion of fossil fuels are a major source of airborne metal contaminants in natural waters.

Agricultural runoff containing trace metals from animal and plant residues, phosphatic fertilizers, specific herbicides and sewage nutrients also contribute to metal enrichment.

REFERENCES

Schwoerbel J. (1987). Handbook of Limnology. Ellis Horwood Ltd. West Sussex (5th edition).

Connell D.W., Miller G.J. (1984). Chemistry and Ecotoxicology of Pollution. John Wiley & Sons, Canada.

Goldman C.R., Horne, A.J. (1983). Limnology. McGraw-Hill Book Company, Japan.

Cole, G.A. (1979). Textbook of Limnology. 2nd edition. C.V. Mosby Company, London.

16

BAPTISM BY FIRE

Humans have always been interested in future - most are even obsessed with it. Attempts to forecast, or read 'the fate' have been made since time immemorial. But the science of futurology gained a major impactus when, during the late 1960s the world was made aware of the environmental perils that loomed ahead. Dennis Meadow's *Limits to Growth* - which was an study in futurology in the sense that it tried to forecast the state of the world if development and pollution continued to grow at the rate prevailing in the 1960s - played a major role in sensitizing people around the world towards environmental pollution and needs for conservation. Now, too, exercises in futurology are often excercises in balancing developmental compulsions with environmental health.

I had my initiation into futurology in a rather dramatic manner.

Futurologists say that we should study future and be prepared for it lest it takes us by surprise. As if to make me ever remember this adage my initiation to futurology took place with quite a surprise.

That was 1976. In those days I was teaching pollution control in the Department of Metallurgical Engineering at IIT Bombay. One sultry Mumbai (then Bombay) afternoon, taking in my share of air pollution, I received a communication saying that I have been nominated as a member of a task force on futurology set up by IIT under the convenership of eminent hydrologist and engineer-philosopher Prof J.T. Panikar.

I was not only surprised on receiving this note but shaken as well because I knew nothing about futurology and it was never good form to be on an any academic platform at IIT without being well-groomed. I therefore hastened to ring up Prof Panicker and blurted out to him, 'Sir, what can I do on your task force because I do not know futurology from Adam'. Prof Panickar calmly replied 'so dont I ... it is a new science and nobody is trained in it ... thats why we have set up this task force to chalk out where and how to start'. I said, 'pray can I ask you what is my qualification for this job'. He said 'you are a writer so we trust you have more imagination than average'.

In those days I *was* a writer of sorts. It is another matter that even in those days I was as confused about my priorities as I am today; as a consequence I was never able to make up my mind whether to specialise in fiction or non-fiction, prose or poetry. So I wrote in all these forms. Nevertheless, inspite of this lack of focus, I *had* managed a reasonably consistant run of success and was getting published regularly in the popular magazines of the times. Some of my poems had also been set to music and were being sung, off and on, during cultural functions. This added - though unfairly - to my image as a 'writer'.

I had also suspected, and was later found to be correct, that my nomination to the task force had been engineered through subtle lobbying of dear friend Dr K.K.S.Bhatia. He was working with Prof Panikar as Research Engineer (and now is Scientist-F and Coordinator of divisions at National Institute of Hydrology, Roorkee) and never lost any opportunity to project my credentials before anyone who was prepared to listen. It was Dr Bhatia who had told Prof Panikar about my exploits as a writer - undoubtedly with a doting brother's natural inclination to exaggerate.

So there was I, a week later, sitting in Prof Panickar's room with other members of the 5-strong task-force. There was Prof Rao, Head of Electrical Engineering Department, who was known for his command over English language in a set-up where the average standard of English language - as indeed of everything else - was exceptionally high. Then there was Amitava Gupta - a handsome though reticent lecturer in the Department of Humanities, with Tufts University (MA) and

University of Georgia (PhD) behind him. Then there was V.N.Mishra - Librarian - who always exuded so much energy that it always looked as if he is coming direct from gymnasium after a swell work-out. Indeed the entire Central Library of IIT Bombay seemed to pulsate with energy ever since he took over its librarianship. Chairing the group was Prof J.T.Panickar. A civil engineer by training he was as much known for his contributions to water resource engineering as for his ability to integrate the earthy issues of engineering with an artist's subjectivity and a philosopher's human - oriented thinking.

In that and subsequent .meetings various options were explored - starting an MTech programme in futurology, introducing courses on forecasting in the BTech programmes, taking up multi-departmental and cross-departmental research etc etc. After a while it was decided to make a small and 'safe' beginning by organising a national workshop on futurology.

In due season the workshop was flagged off. Delivering a crisp and stylised keynote address was a tall, well built, man with a long though shapely nose and dreamy eyes. He spoke with an imperiousness - cultivated as I realised much later - as if he was the Lord of all he surveyed; as if futurology was the most important things to have happened to India, as if he was sought after and consulted by everyone who mattered, most of all by the then Prime Minister Smt Indira Gandhi. Those were the times when India was under internal emergency. The air used to be thick with scent of intrigues and coups. A veil of secrecy had enveloped the corridores of power. No one knew when he or she would fall from grace and be put behind bars. In such an atmosphere Dr Seth's name dropping and pronouncements often had a chilling impact. When he was not sending shivers up your spine, he was captivating you with his one-liners. He had brought with him copies of a collection of short stories he had published some years back. With deliberate casualness, he had let the book be circulated among the IIT faculty. The stories contained in the book had nothing to do with futurology but letting them be seen around helped Dr Seth to deepen the aura he had so carefully built around him. It helped in projecting him as a *virtuouso par excellance*; an omnipotent guy who had the spine of a leader, the brain of an academic, and the heart of a writer! He also spoke in underplayed tones, but at the most opportune moments, of the time he used to open the bowling for Agra University and had to decline an invitation to play in Ranji Trophy simply because the siren call of University of Manchester (where he did his second doctorate) had proved too enchanting!

Everything was, of course, a great put-on because Dr Seth was, and has remained, a man essentially simple and sincere at heart. He *did* have lots of talent. A great deal of what he directly or indirectly said about himself *was* true. But in those days he had decided to project not simplicity but infinite virtue and complexity. It had a purpose, and the purpose was well served - this book is one of the many many examples of the results that were eventually achieved by the impact Dr Seth generated through his personality in those days. No one knew what the hell futurology was but everyone *knew* (or thought he/she knew) Seth was! And everyone thought if futurology is something as virtuous, as stylish, as enchanting, as diverting as this man is, it must be something worth pursuing!

The workshop ran for ten days. Speakers from all over India laboured to find their way through the maze of confusion that always surrounds any new multi-disciplinary branch of knowledge. Was futurology a *science*? Or an *art*? Was it something *really new* or just an old goat in a new awtar? Did it have staying power or was it going to be just a fad?

Seth was there all through. During question -answer sessions after the speakers' lectures, he often clicked his fingers to call for the mobile mike and answer questions from the audience which were supposed to be answered by the speaker. Sometime after an speaker had answered a question, Seth answered it *again*. He seemed to be in total command. He seemed to know everything!

A couple of days after the inauguration of the workshop I received a phone call from Mrs sudha Chopra, producer from Bombay TV. 'We want you to set up a 30 minute programme on futurology for us, will you?' she asked. 'Sure' said I. Farooq Sheikh, who later got a good spell of success as a hero of silver jubliee hit films like *Katha* was to be the anchor. I was to help DD in recording mini-interviews with Dr Seth, and Dr De (the then Director, IIT Bombay) and was later to participate in a discussion along with Dr Rashmi Mayur on

TV. 'why not' said Sudha Chopra, 'I bring my crew to IIT and record impressions of some of the students as well?'. 'Capital' I said.

Sudha Chopra was a fair, handsome, and stately lady in her thirties. She had an air of class and efficiency about her. She treated me with a mixture of courtesy, affection, and authority that remainded me of elder sisters or young aunts. We got along very well and did quite a few programmes in tendum.

Television was a novelty in those days. Only Delhi and Bombay had TV centres and their transmissions - black and white as they were - were restricted to a radius of 50 km from the transmission tower. Setallite and colour was still years away; multi-channel transmissions yet more years away.

I went to meet Dr seth one evening to discuss the broad structure of the programme and fix up, on behalf of Sudha Chopra, a time for recording his interview for the programme.

He was in his suit at the IIT Guest House, with a bottle of whisky, a few bottles of soda, a flask of ice cubes and a glass in front of him. After my arrival he brought another glass and asked me, 'should I add some soda or would you like it neat?'

'Thank you I said', defensively, almost guiltily, 'I do not drink'!

'Make a beginning now', he said, 'you don't know what you have been missing so far'. He then poured some whisky in the glass he had brought for me, mixed some soda, put an ice cube, and handed it to me. He then lifted his own glass and held it in front of me saying 'cheers'.

I gingerly lifted my own glass and touched it with his, saying 'cheers' with a lump in my throat.

Thirteen years later, in 1989, when he had come to Pondicherry to flag off the workshop-cum-seminar on futurology I had organised, a similar scene was enacted when I went to his hotel room to meet him after his arrival from Delhi. When he raised his glass for the toast, I said 'You may not remember it Dr Seth but it was you who initiated me to this path of sin'. I then told him about that 1976 evening at IIT Guest House. He bellowed with mirth and said, 'then you must be a seasoned sinner by now'.

Well I wasn't really. Not because of some steely self-control but because liquor never seemed to give me any 'kick', however much I drank. When I told him this he said, 'come on, you dont seem to have drank enough. Now, take this full bottle with my compliments and clean it up. Then tell me if you dont get a kick!'.

I gently reminded him that we are together for a bit of business called workshop-cum-seminar on futurology and my endevours to get a kick may not exactly agree with that little hitch.

He saw the point and did not press me further.

Memories, memories, memories! After Dr Seth had recorded his interview and flown back to Delhi, and after Dr De's and the student's mini-interviews had also been 'canned', Sudha Chopra rang me up to tell me that the programme, anchored by Farooq Sheikh, which would mainly comprise of discussion on futurology between myself and Dr Rashmi Mayur, with excerpts from the recorded mini-interviews thrown in here and there, is scheduled for 17 April.

'Very good', I said, 'when should I report for recording of my part'.

'Oh dont worry about that' she said, casually, 'we will have that on the spot'.

'You will record it before the telecast?' I said, becoming a little apprehensive.

'Sort of,' she said, even more casually, 'we shall record it as it is telecast'.

'Whaaat', I screamed, now fearing the worst, 'Do you mean to say it shall be telecast live?'

'Yeah. That way you would be saved the trouble of coming for recording'.

'But', I wailed 'that is no trouble for me ... no trouble at all ... you kindly record the programme'.

'Cant', she said, 'it is all fixed up already ... so do make sure to be there by 7.30 as we go on air sharp at 8; next to the Marhati news'.

'Now look Madam ... Sudha*ji* ... please ... dont do that to *me* ... you know ... record the programme please ... if you want

I will take special permission from PV (P.Venkatamurthy, the then Station Director)'.

'I wish it were possible. Now Abbasi*ji* dont think so much about it. We will have you live. It won't be any problem'.

'It will be, it *will* be' I screamed, 'in the live telecast there is no room to rectify if I make a mistake ... suppose I goof? Everyone will see it'.

'But you won't goof', she said, sweetly yet raising her voice slightly and speaking faster to smother my attempts to interrupt her, 'Why do you think you will goof? You won't. Now don't bother about it anymore. See you on 17th. Make sure to be there by 7.30. It takes 10 minutes for make-up and then some time will be needed by Farooq to organise the programme ... yes yes, no no ... now dont bother about it at all ...'

But I *was* bothered. Bothered as hell. If it were recitation of poems, or a play, or a game of cricket I wont have worried about live telecast. When you recite a poem or participate in a play you know your lines and only have to make sure that you dont forget them at the crucial moment. Little possibility of making an ass of yourself. While playing cricket (or tennis) you can hit a stupid shot but *that* is not half as shameful for an IITian as making an intellectual error. While being interviewed live, not knowing what question will be hurled at you when, there is all the possibility of making such an error. So there *were* grave risks in letting yourself be telecast live in this case. The whole of Bombay and the whole of Poona will be watching. The whole damn IIT will be watching. And with arc lights blinding you, camera whirring at your face, Farooq Sheikh plastering you with questions, and the thought that two crore pairs of eyes are watching your every move, hanging on to every sound you utter, live telecast was not going to be the cakewalk Sudha Chopra wanted me to believe.

I wasn't as jittery about goofing in front of the rest of the world as I was in front of my IIT pals. The chaps with whom I moved day in and day out. If I goof they will make sure that I am reminded of that at all times. I wont ever be able to look them squarely in their eyes again. Life will be one long, dark shadow.

The thought of what had happened to PC (Pradip Chopra) deepened my depression. He was on a quiz programme and when answering a question correctly as 'schizophrenia', had a slip of tongue and pronounced the word 'schizophrenalia'. That was the end of him. He was ribbed and ribbed and ribbed. Ribbed by so many, so often, so much that he lost weight, developed dark patches below his eyes, grew beard, and became unrecognisable from the bright and bubbly PC that he was before the catastrophe.

I rang up Sudha Chopra to make one last, desparate plea. But was told that she has gone off for a recording.

The information she had given, that Farooq Sheikh will 'organise' the programme between 7.40 (after my make-up) and 8 (before we go on air), was even less reassuring than the prospect of the live telecast. At IIT the brightest of Chaps had spent ten days hearing lectures of *Creme de la creme* during our workshop and still seemed cross eyed when asked 'what is futurology'. And here a fellow will *anchor* a programme on futurology on the basis of inputs he would receive in a matter of 20 minutes. It would have looked plausible if the fellow was a Harvard topper. But Farooq Sheikh - bright no doubt - was *not* a

Harvard topper. Indeed he earned his daily bread by virtue of his looks rather than his brains.

I rang up Sudha again but she was not available. I knew, then, that the die was cast.

When I reached Bombay TV at Worli on the evening of the telecast I had dark thoughts about Sudha Chopra. But for the lady's stubborn insistence on not recording the programme earlier I would, this time, be sitting in my hostel launge in front of the TV with a bottle of chilled coke, surrounded by my pals. Everyone would be waiting to see me on TV. I would have been the toast of the evening, without a care in the world.

Now also everyone *else* would be in that launge, waiting for my show. The whole scenario would be similarly cozy except for me. Instead of being on air, I am on fire. All because of this lady.

As soon as she saw me, Sudha Chopra complimented me on the choice of my shirt. 'Nice of you to have remembered to come in this striped blue shirt' she said, 'many artists don't shake off their habit of coming in white which causes glare on the TV screen. But in your case it is fine ... we wont have to worry about changing it'. And she smiled sweetly, all at ease.

So the good leady has no more worries, I told myself gritting my teeth, I seem to have taken the last load off her chest.

She escorted me to make-up room, saying, 'Farooq and Dr Mayur are expected any moment ... till then let them finish your make-up'.

The make-up room comprised of rows of cushy chairs with adjustable head-rests of the type one seen in hair-cutting saloons.

The make-up men seemed to be even more relaxed then Sudha. They prattled around making gentle wise-cracks. While they applied pancake or powder-puff with one hand they gesticulated with the other giving their judgements on exactly what Naik (the then Chief Minister of Maharashtra) aught to have or aught not to have done with Datta Samant or Bal Thackerey.

Indeed in the entire TV centre everybody seemed to be toally relaxed except me.

The make-up began. My make-up man placed my head on the head-rest at an angle convenient to *him* and began working on my face as he giggled at a one-liner some other make-up man had shot just then. I felt like a lamb being made ready for slaughter.

It was 7.45 when Farooq Sheikh arrived. Dr Mayur had already come. As soon as Farooq lowered himself on a chair Sudha pulled her own chair near him and began briefing him about the programme in rapid undertones while he kept nodding. She showed him the stills that were to be flashed on the screen, briefed him about the interviews she had recorded, and then pointed towards myself and Dr Mayur. Farooq asked us about our respective academic backgrounds. To me he said simply, '*Dr Sahab* I shall ask you questions about futurology and environment'. He gave a similarly crisp brief to Dr Mayur. Then he turned to Sudha and said in his modulated voice 'All set now; lets move to the studios'.

It was not yet 7.55. Farooq had taken less than 10 minutes to grasp the programme and 'organise' it. I was amazed. Inspite of my preoccupations with my own discomfiture and the apprehensions of the impending disaster, I *was* amazed.

We were taken to a large sound-proof hall containing several 'floors' as studios and were made to sit in one such studio in an order dictated by Sudha chopra. One could see the Marathi news reader doing her stuff on a nearby platform. A TV monitor kept in front of us but out of the view of the cameras was showing us the telecast.

I felt a twang of jealousy for the news-reader. All she had to do was read out from a script. No wonder she seemed to be on a song. Not for her the tensions I was made to endure; of wondering what questions Farooq may field and whether any words will ever come out from my mouth once the cameras ping on me.

The technicians on our floor had barely indicated 'all set' when the Marhati news got over and the announcer announced our programme.

I was looking at the monitor when, from somewhere near the cameras, Sudha frantically indicated me to look straight.

The announcement was over, Sudha waived to Farooq signalling him to begin.

Farooq introduced the audience to the theme of the programme and what they would be seeing in the next half-an-hour. He spoke confidently and fluently; not like a man who had spent all of mine minutes over his homework.

Dr Mayur being senior, was given first chance to answer Farooq's questions. I looked at Mayur gesticulating to emphasise his points and then yielded to the temptation of having a long look at the monitor to see how he was coming on the screen.

Then Farooq said something like - 'we also have Dr Abbasi, a young lecturer and researcher from IIT ...' I stole a glance at the monitor to see myself looking sideways while my name flashed on the screen, superimposed on my own image. At the same time I felt as if the tension inside me is about to make me explode with a bang.

I hurriedly looked back at Farooq who was asking me, 'Dr Abbasi; you are trained as an environmentalist. Then what has made you interested in Futurology?'.

The world now looked at me, expecting my answer. The moment of truth had come. I suddenly found myself totally cool. Totally in command. So totally that I felt as if I can doze off it I wanted to. Farooq had asked a question. I was answering it. Nothing else seemed to exist. Nothing else seemed to matter. Then Farooq asked some other question, then some other. Then yet another ...

Half an hour later when it was all over and as soon as she had bid goodbye to Dr Mayur, Sudha turned to me and said, 'You spoke so well ... that other man was all in knots but you saved the programme'.

I wasn't elated because I did not quite believe that she meant it. I thought she was trying to keep up my morale and perhaps compensate me for the ordeal I had been subjected to by her insistence on a live telecast. I wasn't going to breath easy till I had found out how IIT Bombay had thought of performance!

Well my pals, for once, were gracious when I ran into them. I waited for their solvos on the slips I might have made but no solvo ever came. Rather they seemed to be, for once, lavish in their praise. It *did* seem as if I had survived my first live telecast!

Memories, memories, memories. Of meeting Daadu (Dr Mahesh Chattopadhyaya) after a gap of 15 years when he came from Benaras ostensibly to attend my workshop-cum-seminar on Futurology in 1989 but actually only to meet me. Of meeting Sen, my classmate and fellow-gangster of the sixties, who had by now been transformed to an incredibly distinguised Professor Kalidas Sen. Of long evenings we spent together looking back. Talking of Futurology during the seminar but looking back whenever we got together. Looking farther and farther back!

Memories ... of the great pleasure of working with Naseema who, finding me short of staff in those days, took upon herself all the roles ranging from the top executive to the ad-hoc labourer, including steno-typist, treasurer, manager, caterer, on-line trouble shooter ... switching from one job to other with amazing dexterity.

The phone is a-twitter. I am shaken out of my transe and forced to look ahead. The clock chimes. Time is passing; its onward march relentless as ever. Arya looks in 'Sir, we have to reply to the editor of *Environmental Software*'. The phone rings again; Dr Natarajan's resonant voice comes on the line, 'Professor *Saheb*, Government of Pondicherry Officers are asking for the report ... when can you submit it?' Prabhu comes in and hands me a telegram from Er Sivarami Reddy, which says 'arriving on twentieth'.

Tasks, deadlines, commitments. So it is future time again. Back to the future!

17

HOW I WROTE ENVIRONMENTAL IMPACT OF WATER RESOURCES PROJECTS

It was on a balmy March 1982 afternoon, while I was travelling with Dr S.Maudgal - then Director, now Advisor in Ministry of Environment and Forests, Government of India - that the first thoughts occurred which ultimately led to the writing of the book *Environmental Impact of Water Resources Projects.* The thoughts came as uncertainly and vaguely as the first thoughts generally occur. We were on our way from Trichur (now Thirussur) to Piravom, heading for the up-and-coming Muvattupuzha dam. Dr Maudgal was in Kerala as the convenor of the ministry's Environmental Impact Assessment panel and I was with him as one of the panel members. A day before we had inspected the partly constructed Chimony dam near Trichur and had then talked late into the night sharing our concerns at the monumental importance of assessing the impacts of new water resources projects and the equally great but forbidding lack of know-how to go about the task. Of course we knew of the studies done on the impacts of dams across the world - High Aswan dam, Kariba dam, Hoover dam ... but the question was - how much of that knowledge is relevant to the dams in India? We thought of Hirakud dam across Mahanadi which was founded in 1946; of Bhakra which came into existence a few years later across Beas. We knew that there were well over 1000 dams in existence ... surely there must be a lot of studies done on these dams here and there by universities, research stations, PWD and pollution boards. But where was all that information?

Now, on the road, as the coconut trees and paddy fields whizzed by the car, we were mulling over the previous night's discussion when Dr Maudgal suggested that as a small first step I contact universities, research stations and dam authorities in Kerala and let him know of whatever comprehensive studies there were. As I was then with Centre for Water Resources, Kozhikode, the task appeared rather simple to me and I readily agreed.

As soon as Dr Maudgal's visit was over, I set about the task with the naive's enthusiasm.

And soon I was getting my first disappointments. Whomsoever I was approaching was telling me there are some very comprehensive studies though we have not done any ... why don't you contact so and so? Indeed everyone had the 'general' information of a speculative nature on the ills of large dams but where was the specific, quantitative and comprehensive study we were looking for? It simply wasn't there.

That was 1982. Later, I am sure, institutions such as Centre for Earth Science Studies, Thiruvananthapuram, and Kerala Forest Research Institute. Peechi, have completed rather thorough studies on Idukki and a few other reservoirs. We ourselves did an extensive study of Kuttiadi dam during 1984-87. But there was little in 1982 save bits and pieces here and there.

A few months later, after a discussion with Dr Maudgal at his office in New Delhi, I formulated a project proposal to study three mature, three just commissioned, and three due-to-be-constructed dams, representative in size, elevation, and environment. 'Total' impact studies were envisaged, covering all aspects from A to Z. The project was estimated to cost Rs 29 lakhs (equivalent to more than double this figure at today's value of rupee). It was an ambitious proposal at least in terms of the envisaged coverage but it never came to be funded as it ran into one unexpected block after the other. By and by so much time passed that we lost track of it. The project is yet to be rejected but-more significantly-is yet to be funded. In July 1987 I moved to Pondicherry. The Madras editions of newspapers - to which I had naturally switched from Cochin editions - almost every day carried news on 'Mettur dam level'. During the later part of 1987 the nation was going through a drought and Mettur was even more prominently in the news than it normally is. Even though Pondicherry city was not getting the roughest end of the drought like the neighbouring Tamil Nadu, the drought was nevertheless very much in everyone's consciousness and the daily solicitude involving Mettur not only brought back the memories of 1982 but sensitized me more and more to once again pursue what was left out then.

A few telephonic conversations with Dr Maudgal followed. Soon I had submitted a proposal to survey the water resources projects in India and prepare a state-of-the-art report on the environmental impact assessment of those projects. After wetting the proposal Dr Maudgal advised me to scale down my ambitions and study the Krishna, Mahanadi, and Godavari basins first before taking on all-India. As the later events proved, this advice saved me from ending up a wreck.

The project was sanctioned during mid-1988. Ironically the drought which was the prime impetus for the project also caused a drag in its funding because in view of the drought, the Government of India had put an embargo on all new sanctions sometime during early September of 1987. Only when the embargo was lifted half-an-year or so later, the project was sanctioned.

I hadn't become wiser from my 1982 experiences with the Kerala-based projects. I was as breezy and cocky when I began this survey as I was then. The fact that Dr Soni (and later Ms Tessy) was there to assist me, gave an added fillip. My assessment was that lots of water had flown through the sluices since 1982. The environmental movement had spread far and wide. By now, surely, there have to be a plethora of information ready to be tapped.

Well, it wasn't. Not in any case as 'ready' as I had reckoned. There were solid walls and pitfalls galore into which the garden paths kept leading me, but to few doors. Indeed, like before, everyone knew there was a lot of information and the next person was bound to have it. The next person was equally sure ... and so on. Often the chain finally brought us back to the point at which we had begun!

Of course, slowly - almost imperceptibly - the information was trickling in. By the time we had scanned all possible primary, secondary and tertiary sources - spanning books and journals, universities, government agencies such as public works/irrigation/electricity departments, dam authorities and pollution control boards, and non-governmental organisation - we had come to the end of 1990 and a substantial document had taken shape.

In early 1991 Discovery Publishing House saw the manuscript and immediately decided to publish it!

Curiously, more than five years after it was published, the book remains one of the very texts published on this subject - a very surprising fact considering the overwhelming importance of the subject.